I0071600

The Archaeopteryx Controversy

By Richard B. Pittack, B.A., M. Div.

The Archaeopteryx Controversy
By Richard B. Pittack

Copyright © 2007. Richard B Pittack. All rights reserved.

Published by Walden's Computer Services

Printing History:
September 2007, second edition.

Title: The Archaeopteryx Controversy
ID: 1052500
ISBN: 978-0-6151-8232-2

CONTENTS

THE *ARCHAEOPTERYX* CONTROVERSY

The Evidence for Evolution Is *Not* Overwhelming

By

Richard B. Pittack, B.A., M. Divinity

A Treatise on the Authenticity of the *Archaeopteryx* Fossils and Their
Role in the Paleontology and Geological Controversy Between
Evolution and Creationism

"The LORD .... Heals the brokenhearted
and binds up their wounds. He determines
the number of the stars and calls them each
by name. Great is our LORD and mighty in
power; his understanding has no limit."
Psalm 147:3-5

NIV

Scriptural quotations used in this book are from the Holy Bible, New International Version (NIV), Pocket – Size Edition, Broadway & Holman Publishers, Nashville, Tennessee, copyright © 1973, 1978, 1984 International Bible Society.

INTRODUCTION

Dear Readers,

I never dreamed that I would produce an entire book based on the controversy that surrounds "the most beautiful fossil in the world" – *Archaeopteryx*. I have been thinking of the unusual circumstances that led me to write of this strange creature. It all goes back to a family incident occurring fifteen years ago involving my wife, Kathleen and my son, Kyle.

Becoming interested in dinosaurs; a large collection of books began to accumulate in my personal library. Making certain of acquiring only the best, available information; the acquisition of such knowledge grew to approximately one-hundred books. Of course, these were purchased over a long period of time and at least 40% of them were juvenile books. If the juvenile books are included, the collection of books is well over a hundred.

I suppose my enthusiasm for dinosaurs engendered the same interest in Kyle and after a short time, he was fairly adept in dinosaur craft and able to name and describe the various species of the common and not-so-common dinosaurs. I had procured a set of 28 books for Kyle. There came a day when Kathleen was perusing one of the dinosaur books. She suddenly exclaimed with delight, "What a beautiful dinosaur!" Kyle immediately took the book in order to see the source of his mother's remark. With the proper enunciation, pronunciation, and speech articulation, Kyle blurted out with some aggravation, "This is not a dinosaur; *Archaeopteryx* is a bird." How was Kathleen to know that an ancient bird would be the focus of attention at the same time studies having to do with dinosaurs were the main issue? This logical but technical slip-up was not to be tolerated by a three year old scholar even though Kathleen was that scholar's very own mother.

However, according to the convictions of evolutionists, Kyle and Kathleen both would have been correct in their assessment and scientific evaluation of *Archaeopteryx* since some evolutionists believe *Archaeopteryx* to be a bird and other evolutionists believe it to be a dinosaur. And so, a family incident became a harbinger of what would take place in my life, fifteen years later; the writing of a book that brings to the center of our attention one of the issues in the evolutionism-creationism debate – was

Archaeopteryx a full-pledged bird or was it a lizard or dinosaur? I will quote some remarks from chapter two of my first book – "Was Darwin Wrong? YES":

"Science is not a game to be won by the side that can set forth the best arguments. Rather, it is a tool to assist men in their search for truth in the natural world. The search for truth is serious business and science is not an entertainment specialty designed for those individuals who love to engage in heated debates. Science has a greater purpose than to glorify those individuals who are the most skilled and proficient in expressing either the philosophy of creationism or the philosophy of evolution."

P.15

In the mind of the creationist, the importance of winning the argument that the *Archaeopteryx* is a bird rather than a lizard or feathered dinosaur is not so much the issue. There is a higher issue involved than the mere identification of the biological identity of the *Archaeopteryx*. The central issue of God's existence is at stake. Did God, in fact, create *Archaeopteryx* or was it the product of the "holy trinity"; blind chance, spontaneous generation, and evolution (Transformation of species).

The evolutionists often criticize creationists since the doctrine of God creating species *ex nihilo* (out of nothing) strains credulity. My thirteen year old granddaughter, Candace, brought a school book home describing the origin of species. Against the slow methodical transformation of organisms through the course of evolution, the book gave the idea that the biblical Creation was some type of magic – "Poof", everything appeared. However, God did not create by slight of hand or by trick mirrors.

In creationism or evolutionism; what following concepts strain the imagination more and appear to be less logical and push credulity to unprecedented limits? Very briefly, let us follow these two concepts!

CREATIONISM which has *ex nihilo* (out of nothing) at the center of its doctrine; the following texts are indicative of the Biblical explanations for the universe and origin of life:

- GENESIS 1:1 "In the beginning God created the heavens and the earth."

- PSALM 33:8-9 "Let all the earth fear the LORD; let all the people of the world revere him. For he spoke, and it came to be; he commanded, and it stood firm."

- HEBREWS 11:3 "By faith we understand that the universe was formed at God's command, so that what is seen was not made out of what was visible."

[Scriptures from the NIV]

EVOLUTIONISM; the doctrine has the following explanations for the universe and origin of life:

- "Nobody times nothing equals everything" (John MacArthur, a creationist).

- Darwin + Lyell = EVOLUTION or

- Natural Selection + Deep Time (millions and billions of years) = *anything is possible* even SPONTANEOUS GENERATION AND EVOLUTION.   [Thomas Huxley who coined the word "agnostic" wrote in his *Discourses Biological and Geological*, Pp. 256,257:

" .... I should expect to be a witness to the evolution of living protoplasm from not living matter ....That is the expectation to which a logical reasoning leads me: but I beg of you once more to recollect that I have no right to call my opinion anything but an act of philosophical faith."]

The following words keep echoing in my mind:

"'For my thoughts are not your thoughts, neither are your ways my ways, declares the LORD. As the heavens are higher than the earth, so are my ways higher than your ways and my thoughts than your thoughts.'"

Isaiah 55:8-9

NIV

Evolution demands faith to uphold its scientific (?) claims. What was Huxley's admission? ; Life springing from non-life has no scientific evidence to support it. I appreciate the legacy of his candid and open remarks but I believe that it was only because he could not possibly have written anything else, which would have been considered sensible.

- Quantum mechanics = Quantum Fluctuations as being the CAUSE for generating the EFFECTS seen in a hundred billion galaxies.

Let us go back to creationism!

G.K. Chesterton's statement will conclude this short section on God's existence as Creator. It is my favorite quote:

"It is absurd for the evolutionist to complain that it is unthinkable for an admittedly unthinkable God to make everything out of nothing, and then pretend that it is more thinkable that nothing should turn itself into everything."

[In conjunction with the above thoughts, my readers might want to read the last two appendixes even prior to reading this book. They are entitled "A Spider's Web" and "The Steady State Theory"]

This book has been written for those young people who, in the course of their schooling, have had a steady diet of evolution and need another view; for those people who, in their twilight years, became convinced that evolutionary science has made the existence of God and the credibility of the Bible obsolete and mundane. This book will help such readers to see there is another side to the issue of origin of species. Evolution is not the only teaching which attempts to explain life in the universe. In fact, it has no answer for the start of any life forms.

Few people like controversy. Even the Bible warns against needless arguments stirred up by trouble makers and engaged in by men for the simple act of arousing divisions and factions (Acts 26:3; I Timothy 1:4, 6:4; Titus 3:9). But this warning is not to be understood as applying to those of us who are striving to meet the great controversies that affect society and our own personal life.

Creation is a proposition, which needs to be considered as a proper alternative to evolution in spite of what evolutionists say about the term "creation-scientist" being an oxymoron. This book affords readers the opportunity of comparing the SCIENTIFIC PHILOSOPHY of evolutionism with the SCIENTIFIC PHILOSOPHY of creationism. You are to be the judge of which philosophy is logical and rational; centered in systematic truth. [The *illogical part* of evolution is *not the fact* that kinds of animals and plants can *demonstrate variation of species* but rather to the alleged fact that their existence can be traced back *to non-living matter*, that from a simple (?) cell *all the complexities of life* have arisen or evolved; that the dogmatic assertion of evolution being an unequivocal and irrefutable fact while, by the same token, having *no explanations for its deficiencies* and *not any proof* to back up its declarations]

Evolutionism and Creationism stand as opposing warriors with their arrows and spears pointed at the center of your heart. Evolutionism, I believe, is aiming to deaden your reasoning powers and to remove your fear of God; Creationism is aiming to convict and awaken your heart to the respect and reverence due God who created you. May this book provide you with a shield of protection against the deadly forces of a philosophy aimed at destroying your happiness and right to think clearly; free from the intense barrage of propaganda, which is the liberal and materialistic mind-set of this world's society.

The deadly force of evolution will not go away for creationists by simply denying its existence. As a creationist, I seek to be another "lawyer for God." Not that God needs to be defended by the "closing" words of this book but rather that these words might open the eyes of those of you who are already convinced of the scientific nature and correctness of the evolutionary doctrine. I would urge you to accept a belief that will afford you greater happiness and satisfaction than what the doctrine of evolution could ever hope to inspire or empower.

In *The Archaeopteryx Controversy*, you will see the argument as it unfolds in each chapter; its relevancy to the belief in creation or evolution; its impact on the mind as it helps you to distinguish truth from error; its role in the destabilization of evolutionism and its vindication of creationism.

May this treatise do more than close the mouth of evolution with all its voice in proclaiming that illogical reasoning advanced in the name of science. May it open up the hearts of the readers of this book along with my heart to the feasibility of the Creatorship of God and to the conviction that recognizes Him to be not only the *Creator* of the universe but also the recognition of Him as the *Judge* and *Redeemer* of this planet and our very own life.

Respectfully,

Richard B. Pittack

Palmdale, California

May 10, 2007

CHAPTER 1
THE LIMESTONE TREASURES
(The Discovery of *Archaeopteryx*)

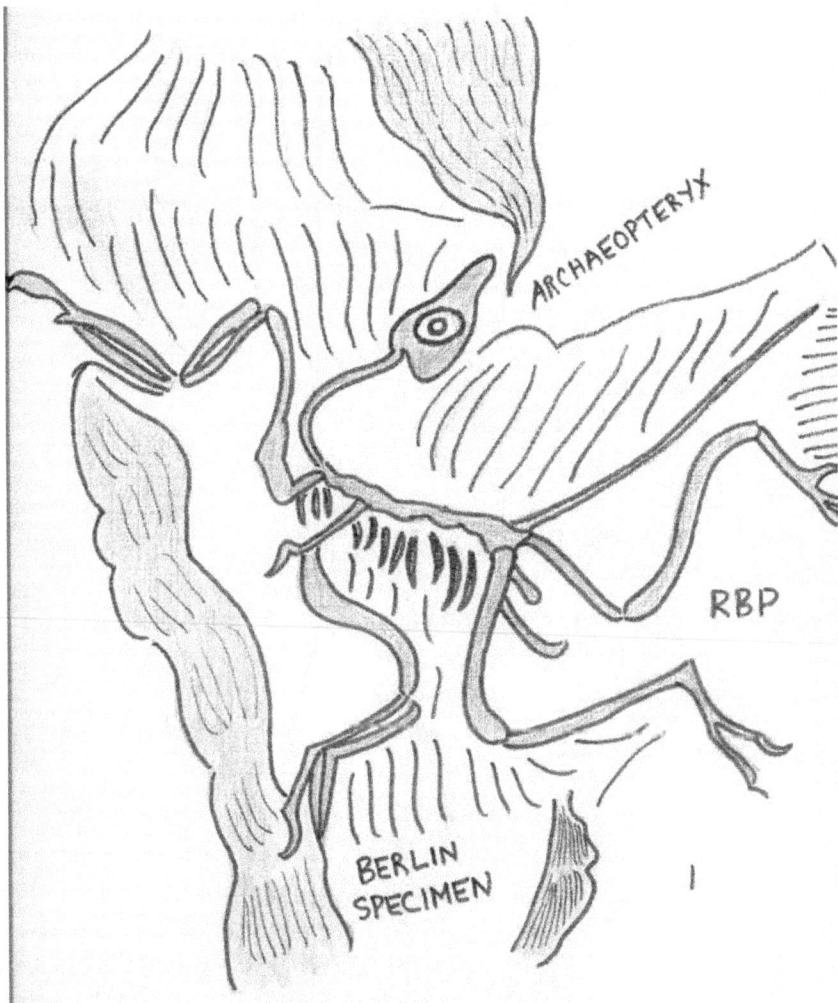

"They will summon peoples to the mountain and there offer sacrifices of righteousness;   they will feast on the abundance of the seas, on the *treasures hidden in the sand.*"

Deuteronomy 33:19

NIV

In 1861 a single, beautiful preserved feather was found in a quarry near Solnhofen. This discovery was made in the rocks of the Upper Jurassic Period and in the greensand. According to evolutionists, it moved the history of birds back well over 100 million years. Since no bird had ever been discovered in limestone beds, some scientists thought the feather to be a hoax. Hermann von Meyer from Frankfurt confirmed the fossil to be genuine. He gave the newly discovered animal a neutral name (neither bird nor reptile): *Archaeopteryx lithographica* (Ancient wing from the lithographic limestone).

Not long afterwards an almost complete feathered skeleton was found 60 feet underground in the Ottmann quarry. This remarkable discovery was made in the bed of an ancient lagoon in Bavaria. Limestone was being quarried by workmen who uncovered a flag stone with the remains of a creature that was birdlike in appearance and about the size of a crow. The fossil had clawed fingers and a long bony tail of reptilian form. But it also had a wishbone and feathered wings resembling modern birds. The feathers were asymmetrical in shape; identical to the feathers of present-day birds of flight.

"In 1861 Hermann von Meyer described [this] fossil that appeared to be intermediate between reptiles and birds ... the fossil had wings and feathers; but it also had teeth (unlike any modern bird), a long lizard-like tail, and claws on its wings."

Icons of Evolution Science or Myth

Jonathan Wells

P.112

In 1862 Richard Owen was superintendent of the British Museum. He sent a representative (George Waterhouse, the Keeper of the Geological Department) to Bavaria to purchase Dr. Karl Haberlein's *Archaeopteryx* which he had put up for sale. Haberlein was the district medical officer for the quarry and obtained the fossil in exchange for the medical fees he was owed. At that time, the specimen cost the equivalent of a small fortune – 700 pounds. But the amount also paid for over 1,000 other fossils from the Solnhofen quarries in southern Germany. The unprecedented price of 700 pounds, at 1862 values, was about $3,500. Owen offered 500 pounds to Haberlein but came up 200 pounds short. Owen met with his trustees and they gave him the needed deficit. According to the author Steve Fiffer, in his book *Tyrannosaurus Sue,* writes in a humorous style aimed at reminding readers of the fossils main character; the British Museum with this purchase "had added another feather to its cap" (P.47).

When Haberlein accepted the offer he, no doubt, suffered sorrow as he watched his prize specimen of *Archaeopteryx* leaving Bavaria and sent to London. But he did take comfort in the fact that he could now give his daughter the 700 pounds he had promised for her dowry. On November 1 the fossil arrived in London without any damage. Owen examined the split slab of lithographic stone and "as early as November 20, Owen reported to the Royal Society that the fossil was unequivocally a bird, though one of a primitive kind."

Dinosaur Hunter

David A.E. Spalding

P.33

This 1861 fossil ended up in the Natural History Museum in London. It became known as the "London specimen" but in 1877 an even more complete sample of *Archaeopteryx* was unearthed. This relic is known as the "Berlin specimen" found near Blumenberg, Germany. The fossil had a complete head but it was badly crushed. This is the most complete and best-preserved sample of the eight known specimens. Wells writes, "It has become familiar to millions of people as the missing link that confirmed Darwin's theory" (*Icons of Evolution*, P.112).

This specimen had the head at the front of a perfectly articulated body and superbly preserved plumage impressions. The detail is much finer than the 1861 specimen. This new fossil confirmed (for many) the reptile-bird nature of the animal since the jaws were not rimed with a bird's beak but rather edged by sharp teeth similar like those of a reptile. Some evolutionists believe that this animal has considerably advanced our knowledge of the link between early birds and dinosaurs. Other scientists still have their doubts concerning this particular identification as *advancement in knowledge*.

The theory of birds being the descendents of dinosaurs is a matter of major dispute between evolutionary scientists. At the April 2000 Florida Symposium on Dinosaur Bird Evolution, while some scientists were lecturing on evolution of birds from dinosaurs, other scientists were passing out buttons **"Birds are NOT dinosaurs."** This issue will be addressed in a coming chapter.

The fossil has become the center of serious disputations but, nevertheless, is the most unusual and valuable fossil ever to have been discovered on this planet. The paleontologists Lowell Dingus and Timothy Rowe declare:

"To museum curators, the name *Archaeopteryx* rings like that of Rembrandt, Stradivarius, or Michelangelo."

The paleontologist Pat Shipman has called the Berlin specimen "more than the world's most beautiful fossil …. [It is] an icon – a holy relic of the past that has become a powerful symbol of the evolutionary process itself. It is the First Bird."

[The Dingus and Rowe quote as well as the Shipman quote are in *Icons of Evolution,* Pp. 113 – 114]

The Solnhofen quarries of Bavaria, Germany, are called the *Lagerstatten,* which is German for "fossil lode" or "storehouse." Because of the unique prehistoric conditions, this storehouse has preserved many animals in fossil form, including the *Arxchaeopteryx* specimens. There are only about 100 fossil sites around the word that have been selected as *Lagerstatten.* The conditions that formed the beautiful fossils at Solnhofen quarries were the tropical climate coupled with the quiet, warm-water lagoon. No fauna on earth has been better studied than the Jurassic Solnhofen limestone. "Quarrymen and amateur collectors have been splitting these Limestone blocks for more than a century. (These uniform, fine grained stones are the mainstay of lithography, and have used, almost exclusively, for all fine prints in this medium ever since the technique was invented at the end of the eighteenth century.")

Wonderful Life

Stephen J. Gould

P.63

For readers who may have an interest in how lithography is achieved, the following paragraph will give you a simple run down:

"Basically the process involves drawing a picture on a polished surface of the stone with a waxed pencil. The block is then submerged in an acid that etches away the surface where it is not protected by the wax. Then the stone is washed and cleaned of the wax. The result is a raised surface where the wax was and a lowered surface, eaten down by the acid, where the wax was not. The stone is then ready for inking, and the image is pressed onto paper. Although it is no longer practical to use stone lithography, some of the most beautiful illustrations ever made of fossil bones were done with this technique. The use of the Solnhofen

Limestone for lithographic plates gave *Archaeopteryx* its specific epithet, *lithographica*."

The Quest for the African Dinosaurs

Louis Jacobs

P.166

A rare geological phenomenon accounted for the remarkable preservation of these exceptional fossils. Sedimentary particles of extremely small size encased the bodies without any succeeding disturbance from nature. The lime subsequently hardened into limestone and preserved in delicate detail the skeletons of the animals contained within it. The reason for the preservation of the feathers, the limestone would retain the shapes and impressions of tissue that would normally be lost. Some of these early birds were in all probability blown into the sea where they drowned.

Thomas E. Svarney and Patricia Barnes-Svarney write:

"Storms would sweep in dead or dying animals from the ocean; dying land creatures either fell into the lagoon, or drifted into it from the shore. Their bodies fell to the bottom of the lagoon and were covered by soft lime mud; little oxygen was present to decompose the organisms. . The ensuing fine limestone rock preserved, in exquisite detail, the remains of over 600 species, including the smallest dinosaur, *Compsognathus*; pterosaurs by the hundreds; numerous insects; and, of course, the remains of *Archaeopteryx lithographica*."

The Handy Dinosaur Answer Book

P.351

Many scientists suggest that *Archaeopteryx* probably swum and fished in the offshore lagoons of the island. Charles Darwin wrote of *Archaeopteryx* in this matter:

" … Prof. Huxley has discovered, and is confirmed by Mr. Cope and others that the Dinosaurians are in many important characters intermediate between certain reptiles and certain birds – the birds referred to being the ostrich-tribe (itself a widely-diffused remnant of a larger group and the *Archaeopteryx*, that strange Secondary bird, with a long lizard-like tail."

The Descent of Man

Pp.522-523

He also wrote in *The Origin of Species*:

" ... Not long ago, paleontologists maintained that the whole class of birds came suddenly into existence during the eocene period; but now we know, on the authority of Professor Owen, that a bird certainly lived during the deposition of the greensand; and still more recently, that strange bird, the *Archaeopteryx*, with a long lizard-liked tail, bearing a pair of feathers on each joint, and with its wings furnished with two free claws, has been discovered in the oolitic states of Solnhofen. Hardly any recent discovery shows more forcibly than this, how little we as yet know of the former inhabitants of the word."

Pp. 250-251

Archaeopteryx was discovered in the Jurassic sediments. The Jurassic rocks derive their name from the Jura Mountains to the west of Switzerland. They are throughout the central part of England and were described by English geologists as Oolites; a name suggested by the oolitic limestones that are common only to England (Price). These are the "oolitic states of Solnhofen" referred to by Darwin. **Many creationists believe that the limestone sediments were laid down by the reduction of the energy of the waves or currents of moving water from the Noachian Flood. The Great winds of Genesis 8:1 was also a contributing factor for these deposited sediments. The vast sedimentary deposits of limestone along the coasts of England can hardly be accounted for by the doctrine of uniformitarianism. Deltas, *lagoons*, beaches, and sandbars were being formed as the Flood was receding. Many creationists believe that most of the dinosaurs were destroyed at the time of the Great Flood including *Compsognathus*. I will not venture a guess why pterosaurs by the hundreds were accumulated within Solenhofen areas but these creatures along with *Archaeopteryx*, numerous insects, and *Compsognathus* perished in the Great Flood prior to its receding.**

Scientists affirm that *Archaeopteryx* was a scavenger and walked the muddy beaches looking for stranded fish, crabs, and ammonites. Still others make the unlikely proposal that it snapped up dragonflies. I say "unlikely" because the *Archaeopteryx* wasn't fast enough to catch dragonflies and the extreme reason that some evolutionists have dreamed up; reptiles developed feathers for the purpose of catching insects. **Finally, the best suggestion noted above is *Archaeopteryx* was a fish-eater. The teeth that terminated in thick, barrel-shaped roots provide evidence for a seafood diet. It was the creationist Sir**

Richard Owen who demonstrated that such teeth were a trade-mark of the fish-eaters.

Archaeopteryx is described in nearly every book that speaks about the history of life and evolution. This fossil is thought to be the perfect example of a transitional link between reptiles and birds. On the one hand, it shows the primitive features of reptiles, such as teeth, claws, and a long bony tail. On the other hand, it shows the advanced traits of birds, such as feathers and a wishbone. Because of these last two characteristics, it is the oldest known bird. This bird is from the geological period known as the Upper (or Late) Jurassic. Evolutionists date this fossil about 150 million years ago and this makes *Archaeopteryx* the earliest undisputed bird

This fossil find of *Archaeopteryx* (from the Greek for "ancient wing") was hailed as a great victory for evolutionists and a knockout punch delivered by avian paleontology against creationism in the 1860's and 1870's. Stephen Jay Gould spoke of the knockout punch is this way:

"… the best possible example of an unanticipated fact burst upon the scene – the discovery of *Archaeopteryx*, not only the oldest bird but also, apparently, beautiful intermediate between reptiles and birds in its retention of teeth, reduced coating of feathers, and basically reptilian anatomy. Score one knockout blow for evolution."

Leonardo's Mountain of Clams and the Diet of Worms

P.113

My fellow creationists often quote Stephen Jay Gould's and Nile Eldredge's reference to *Archaeopteryx* when they wrote "At the higher level of evolutionary transition between basic morphologic designs, gradualism has always been in trouble.…Smooth intermediates between *Bauplane* are almost impossible to construct, even in thought experiments. There is certainly no evidence for them in the fossil record (curious mosaics like *Archaeopteryx* do not count)."

[Stephen Jay Gould and Niles Eldredge, "Punctuated Equilibria: the Tempo and Mode of Evolution Reconsidered," *Paleobiology*, vol.3 (Spring 1977), p.147]

This is taken to mean, by some creationists, that *Arxchaeopteryx* was not being considered as a *link* between reptiles and birds. Hank Hanegraaff (president of the Christian Research Institute) has written:

"A few years after Harvard's Gould ruled out *Archaeopteryx* as a missing link …"

However, sincere as these words may be, they do not express a true fact. Gould and Eldredge **did not rule out the fossil bird Archaeopteryx as a missing link.** But these two evolutionists did claim that there are no obvious **transitions** between classes. A fossil can be intermediate (a matter of morphology) without being transitional (a matter of origin). Some of my fellow creationists define "intermediate" as referring to a fossil that is a direct ancestor of one life form and a direct descendant of another life form. For some evolutionists, direct lineage is not necessary for the definition of an intermediate and even if a transition was found, **could not be confirmed.** But, be careful! As an evolutionist, although Gould does not see *Archaeopteryx* as a transitional form he does, nevertheless, believe that reptiles and birds were blood related. If my fellow creationists care to see a contradiction in Gould's type of reasoning so be it. However, for me, I do not note an inconsistency. [The reason for such confusion in attempting to arrive at a definition of "transition" and "intermediate"; most evolutionists, in their writings, fail to distinguish between the two. No wonder my fellow creationists are confused when Stephen Gould *does make a distinction*! Throughout this book, I do not make such a distinction since most evolutionists speak of *intermediate* and *transitional* as though they were *interchangeable terms*. **To add to the confusion, Gould can not always be counted upon for making a clear cut distinction between intermediates and transitional forms of life as we shall see in Gould's coming statement]**

With this said, I am going to give you a clear example of another statement of Gould completely quoted out of context. It has to do with the same subject – *Archaeopteryx*. Stephen J. Gould accused creationists of misquotes. **A long time ago, I doubted what he said but now my doubting has been dispelled.** Here is the quote taken from *Dinosaurs in a Haystack* and page 360:

"The supposed lack of intermediary forms [Note this: INTERMEDIARY FORMS] in the fossil record remains the fundamental canard of current antievolutionism. Such transitional forms [Note this: TRANSITIONAL FORMS] are sparse, to be sure, and for two sets of good reasons – geological (the gappiness of the fossil record) and biological (the episodic nature of evolutionary change, including patters of punctuated equilibrium, and transition within small populations of limited geographic extent)." [This is a good quote for creationists to use because it makes Gould sound as though he was writing: there are no intermediary or transitional forms in the fossil record or they are, at the most, very sparse. **You would think so but you would be wrong!**] Here is the rest of the quote:

"But paleontologists have discovered *several superb examples of intermediary forms and sequences,* more than enough to convince any fair-minded skeptic about the reality of life's physical genealogy." [Italics mine]

[Ibid. p.360]

Gould goes on to cite many examples and he mentions *Archaeopteryx* as "an initial bird chock-full of reptilian features" and just the fossil Darwin needed to fill in the "spottiness of the fossil record." You can see for yourself, as a reader, what I say is true concerning misquotes. [Pittack's note: A misquote may not only involve a replacement of the original words but lifting the statement out of context or a deletion of those words which precede or follow a given statement that would give sense and meaning to the given statement itself] Before we continue on, just a few words in dealing with this very important issue:

- First of all, in knowing that time is of the essence to creationist writers, we can allow the hurried reading of Gould's writings to the point of being mesmerized with certain portions of such writings and failing to go any further with Gould's entire thought.

- Second, the misquoted statements appear in other books on creation since the one author trusts the previous author to be accurate and this becomes a copy problem right on down the line thus becoming a source of embarrassment to other creationists. This is why authors from the creationist camp, if at all possible, should check on borrowed quotes.

- Third, this is a practice of which the evolutionist camp is not to be found guiltless. I think both sides need to be gracious and not affix "guilty of distortion" to the opposing writer. A writer can make an honest mistake.

However, although creationists have left out the statement, "Superb examples of intermediary forms and sequences, more than enough to convince any fair- minded skeptic about the reality of life's physical genealogy", Gould makes a flat out contradiction. He makes a definite connection between "intermediary" and "genealogy."

First, Gould wants us to understand that fossils can be *intermediates*. Such a state is merely a matter of *morphology* without being transitional – involving the more crucial matter of *origins or genealogy*. In other words, here is a major contradiction because Gould reverses his distinction between intermediate and transitional. On the one hand, he says that **the intermediate state of a fossil is just a matter of morphology.** On the other hand, he says that **the intermediate state of a fossil is a matter of genealogy.** Not only does he say: there are some superb examples of intermediary forms demonstrating the reality of PHYSICAL GENEALOGY but some SEQUENCES form superb examples of PHYSICAL GENEOLOGY. If this is true, then gradualism stands a good chance of being true through its superb examples of sequences. If gradualism is true at times, then there is the possibility it could always be true in spite of the fossil record containing so many gaps. But gradualism is the theory that frustrates Gould's doctrine of punctuated equilibrium. Why would he advocate in any way the feasibility of gradualism or orthogenesis?

Now that I have confused every body, **the bottom line is this – whether the terms *intermediate* and *transitional* are used interchangeably or not, in no way do they convey "LIFE'S PHYSICAL GENEALOGY."** I can agree that Gould has been misquoted (not maliciously since Gould sets up misquotes by his own contradictions) but can never agree that fossils contain the power to open up the mystery of their assumed physical genealogy. Chapters three, four, and five of this book are devoted to the refutation that fossils convey "LIFE'S PHYSICAL GENEALOGY." [The word "genealogy" infers that these so-called intermediate fossils are *ancestors* and, indeed, *progenitors* and thus supposedly blood related. Gould does not accept straight line evolution (except in his unguarded moments of contradiction) but that does not mean that Gould fails to note blood relationships between ancestors in the fossil domain. **He simply believes that the array of animals and their derivations in the fossil record are not to be perceived as coming about by orthogenesis but understood as rising up as a bush or bushes.** This makes perfectly good sense to me (from Gould's standpoint); among other things, he can subsequently explain why certain species can be living contemporaneously even though they are considered, so to speak, as "father" and "son." In other words, ancestors survive after descendants branch off.

And now back to our original subject – *Archaeopteryx* considered by evolutionists to be a link between reptiles and birds. Gould and Eldredge taught Punctuated Equilibria; smooth intermediates were no longer considered to be a problem against their theory. In fact, they were arguing against progressive evolution or orthogenesis (Straight-line evolution). According to their theory of punctuated equilibrium, a species does not arise by a series of progressive steps made by their ancestors undergoing transformation; the species appears all at once and completely formed. These "bursts" of macro evolutionary changes lead to the rise of a new taxon and in this theory, the number of organisms involved in this fortuitous leap would be few. That is, the number of complete formed species would be small and the fossilizations occurring would be smaller yet; this so-called fact would account for the number of gaps in the fossil record. This is different from saying the *Archaeopteryx* was not an intermediate or link to reptiles. The process of equilibrium is a long interval of time called stasis; the punctuation is characterized by rapid evolution.

Being in disagreement with *gradualism* does not indicate that Gould was averse to seeing *Archaeopteryx* as a "beautiful intermediate between reptiles and birds." *Gradualism* is the theory that macroevolution proceeds through the slow and continuous accumulation of small changes to finally reach large changes. This evolutionary theory predicts the geological column bearing fossil forms will demonstrate evidence of one species progressively transformed into another. **Obviously, Gould and Eldredge did not accept the doctrine of gradualism but this does not mean that Darwin was no longer a major part of their thinking. It meant only that Darwin's theory of very gradual evolution is wrong to Gould and Eldredge. They were not (in a strict sense) Darwinians but they did remain firm, dedicated evolutionists as they propagated their theory. These two evolutionary scientists were *not opponents of Darwin* but rather strong supporters of his doctrine of organic evolution and the mechanics of natural selection. They were simply looking for new answers to the age old problems (for evolutionists) of gaps in the fossil record and why life made their appearance in clusters of new taxon at every level of the geological column.**

Gould writes about Jeremy Rifkin (the most vocal opponent of genetic engineering) in this way: "Rifkin displays equally little comprehension of basic arguments about evolutionary geometry. *He thinks that Archaeopteryx has been refuted as an intermediate link between reptiles and birds* because some true birds have been found in rocks of the same age. But evolution is a branching bush, not a ladder. Ancestors survive after descendents branch off." (Emphasis mine)

An Urchin in the Storm

Stephen Jay Gould

P.233

Note that Gould is more than clear in the fact that *Archaeopteryx is an Intermediate link* between reptiles and birds. He explains that evolutionism should not be viewed as a ladder. That is, it should be seen as a bush and not as a series of intermediate steps. **In other words, Gould wanted to account for the fossil record as it truly exists and what he could see; not reading into it, for example, gradualism of species what he could not see. This approach to paleontology was somewhat similar to the creationists. Only, sad to say, rather than accept the evidence as it points to creation, he proposed the evolutionary solution – the real phenomena of nature was the result of the mechanism of evolution itself.** Gould not only advocates that the oldest bird is a link because of its retention of teeth, reduced coating of feathers, and basically reptilian anatomy but because of its brain size – "This intermediate form with feathers and reptilian teeth had a brain that plots right in the middle of the unfilled area between modern reptiles and birds."

Ever Since Darwin

P.189

[Pittack's Note: *Ever Since Darwin* was Stephen Jay Gould's first volume of collected essays. It was written in *1973* and he was an outstanding researcher. Nevertheless, he apparently didn't read the article written by J.H. Jerison entitled "Brain Evolution and *Archaeopteryx*" *Nature*, 219:1381-82 and dated *1968*. A study had been made of the cranial endocast of *Archaeopteryx* that was, in all important respects, essentially avian. It had cerebal hemispheres and cerebellum which demonstrated typical avian characteristics. Nothing is mentioned about the brain being plotted in the middle area of bird and reptile (Gould's contention). It is a known fact that *Archaeopteryx* was capable of powered flight and considered a true bird. The *Archaeopteryx* was not a missing link as we shall discover later in this book]

Obviously, Gould believed that *Archaeopteryx* was a link and to use the argument that this fossil bird was a mosaic and did not

count as a link between reptiles and birds is to quote Gould entirely out of context and to totally misunderstand the importation of his argument. Leaving Gould and *his actual acceptance of the Archaeopteryx as a missing link*, we shall view the results of this kind of "link" thinking of other evolutionists.

Robert T. Bakker, a famous dinosaurologist, makes this sportive remark concerning the "London specimen" of 1861 and creationists:

"Inveterate creationists, then or now, never allow their faith to fall victim to facts, but any careful, unbiased observer, it was clear that the fossil bird from the Age of Reptiles consisted of a genuine missing link between classes." [The "classes" meaning reptiles and birds]

The Dinosaur Heresies

P.302

This type of thinking was only enhanced in 1877 when the Bavarian quarries yielded a second *Archaeopteryx* skeleton. Detailed analysis uncovered "startling" evidence: *Archaeopteryx* had teeth. This fact was considered to be pro-Darwinian evidence. The teeth, along with other anatomical structures, "proved" the creature to be a definite intermediate between reptiles and birds. The first *Archaeopteryx* was discovered just two years after Charles Darwin's publication of "The Origin of Species." Naturalist Thomas Henry Huxley, Darwin's "bulldog," took advantage of the find and hailed it as the perfect example of a transitional form between reptiles and birds. For Huxley, the anatomical features and stratigraphic position of *Archaeopteryx* was incontrovertible proof for a Darwinian origin of birds.

Darwin refers to *Archaeopteryx* in his two main works – *The Origin of Species* (a later edition) and *The Descent of Man*. We have already cited these quotes but let us refresh our thoughts:

"We have seen that the Ornithorhynchus graduates towards reptiles; and Prof. Huxley has discovered, and is confirmed by Mr. Cope and others, that the Dinosaurians are in many important characters intermediate between certain reptiles and certain birds- the birds referred to being the ostrich-tribe (Itself a widely-diffused remnant of a larger group) and the *Archaeopteryx*, that strange Secondary bird, with a long lizard-like tail."

Pp.522-523 (*The Descent of Man*)

"Not long ago, paleontologists maintained that the whole class of birds came suddenly into existence during the Eocene period; but now we know, on the authority of Professor Owen, that a bird certainly lived during the deposition of the upper greensand; and still more recently, that strange bird, the *Archaeopteryx*, with a long lizard-like tail, bearing a pair of feathers on each joint, and with its wings furnished with two free claws, has been discovered in the oolitic states of Solnhofen.

Hardly any recent discovery shows more forcibly than this, how little we as yet know of the former inhabitants of the world."

Pp.250-251 (*The Origin of Species*)

I suppose that Darwin was happy with this find and especially with the knowledge some scientists were already viewing the discovery of this "strange bird" as lending support to **Darwin's theory of gradualism** with *Archaeopteryx* viewed as an intermediate between reptiles and birds.

Before responding to Bakker's factitious remark that creationists "never allow their faith to fall victim to facts," it is necessary to examine charges brought against the authenticity of *Archaeopteryx*. It would be a waste of time to discuss any "facts" concerning "Ancient (archaic) Wing (apteryx)" if the fossils of this creature have already been proved to be fraudulent. Thus, before continuing on, it is imperative to meet the assault by certain creationists upon the veracity of the limestone fossils of Solnhofen.

It is to the credit of creationists that not all of them agree with such a stance. In fact, probably most creationists believe in the genuineness of the fossils. It is not necessary to undermine these fossils that evolutionists consider to be problematic to the theory of creationism …

The truth of their validity should be *accepted*; the so-called truth of their evolutionary implications should be *challenged*.

CHAPTER 2

ASTRONOMERS CHALLENGE PALEONTOLOGISTS

(Are the *Archaeopteryx* Fossils Fake or Genuine?)

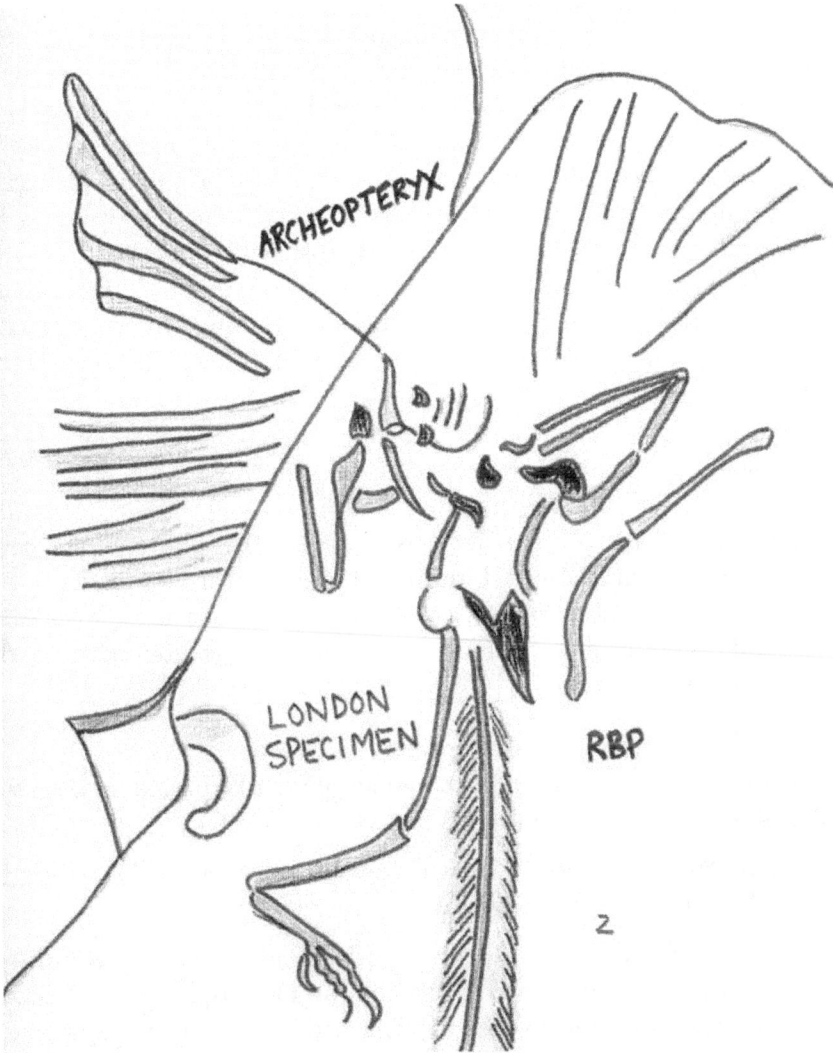

"The priests, the sons of Levi, shall step forward, for the Lord your God has chosen them to minister and to pronounce blessings in the name of the Lord *and to decide all cases of dispute ....*"

Deuteronomy 21: 5

NIV

In 1984 a conference was held attempting to pool together all the new research regarding *Archaeopteryx*. One of the results of this symposium was the serious suggestion that the Berlin and London specimens of *Archaeopteryx* were forgeries. David A. E. Spalding makes this statement:

"This time the suggestion came from a scientist of great eminence, Sir Fred Hoyle, who had previously advanced alternative theories of the origin of life and attempted to discredit evolution. Hoyle is not a paleontologist, or even a biologist, but an astronomer, one of the originators of the steady-state theory of the origin of the universe. He is also a well-known writer of science fiction, and some readers must have wondered in which capacity he approached *Archaeopteryx*. Others involved were Lee Spetner, an American-Israeli physicist, who claimed that a genuine reptile fossil had been faked by putting cement on the rock and adding impressions of feathers. Another was astrophysicist N.C. Wickramasinghe ..."

Dinosaur Hunters

Pp. 41-42

These accusers sought to demonstrate that the photographs of the London specimen showed blurring of the feather impression, a gap between slab and couterslab, and different grain size for the sediments surrounding bones and feathers. This evidence was supposed to show that the fossil was a fake.

[Ibid. P.42]

Gail Vines writes about the part that physicists played in laying the ground work for establishing the *Archaeopteryx* as a fake:

"'One of the world's most famous fossils may be a fake', says Sir Fred Hoyle. 'Any physicist looking at it would have worries.'" ....

"The driving force behind the forgery allegation is Lee Spetner, an Israeli consultant in electronic systems in Rehovot. His suspicions were aroused when he read that the two specimens with the clearest feathers – the London and Berlin fossils – came from the collection of a Bavarian doctor, Dr. Karl Haberlein. Spetner and his colleagues argue that Haberlein gouged out an area around two genuine fossils of a dinosaurlike reptile, and made a matrix with cement which he applied to the fossils. Then, Spetner claims, he preened chicken feathers or the like to the mixture to create the feather impressions...."

"... [Photographs taken by Robert Watkins, of the physics department at Cardiff] show signs of the forger's work: a fine-grained substrate under the feathers, blobs that look like chewing gum that could be the remnants of the forger's cement …. The fossil is split into two halves: an imprint of the creature is reflected into a 'counter slab' created when the rock containing the fossil was split open, like two halves of a mould. Hoyle and his colleagues find elevated or depressed regions on one slab that are not perfectly mirrored of the other. Finally, they point to 'double-strike' impressions of feathers. In a few places the same feather has apparently left two impressions slightly displaced to one side."   "Strange Case of *Archaeopteryx* 'Fraud,'" *New Scientist*, vol. 105 (March 14, 1985), p.3

[Pittack's Note: Lee Spetner had his suspicions aroused when he learned that the two specimens with the clearest feathers came from the collection of the Bavarian doctor Dr. Karl Haberlein. I wonder if Spetner, during his acquired information, also learned about the transaction that took place between the doctor and Richard Owen! Haberlein needed money for his daughter's endowment. He had promised her 700 pounds. Would not Haberlein have fear of being discovered should he have attempted to palm off a faked fossil on the greatest fossil expert in England? Surely, he wasn't that dense as to suppose Owen would not discover, after close examination of the specimen, that he had been duped!

Did Spetner consider the following possibilities? …. … …The original charge was Owen conspired with Haberlein to formulate a plot for the manipulation of the fossil. But this conspiracy could not have been formulated until after Owen's inspection of the fossil itself. Therefore, my original question would make sense; would not Haberlein have fear of being discovered by the greatest fossil expert in England? Also, according to the suspicions of Spetner, the clear feather impressions were already a part of the fossil before *Archaeopteryx* was shipped to Owen. Therefore, I am led to believe that (if Spetner's story of fraud is true) the faked fossil was conceived and initiated by Haberlein. Did he presuppose that Owen would come up with some sort of an anti-evolutionary plan after discovering the fake fossil? Why did Owen being a strict and severe strategist in his approach to paleontology come up with such a nefarious scheme? Wouldn't he fear for his reputation? After all, he made the initial examination of the fossil. What would stop other paleontologists from making their own assessment of the fossil and possibly discovering Owen's original paper, prepared for the Royal Society, was a bogus report based on a fake fossil?]

Fred Hoyle and Chandra Wickramasinghe wrote a small book published in 1986 entitled *Archaeopteryx, the Primordial Bird: A Case of Fossil Forgery* (London: Christopher Davies, 1986).

Ten years later, 1995-1997, the Center for Scientific Creation issued FAQs entitled "What Was *Archaeopteryx?*" Apparently the Center, in its five- paged document, gave credence to the book of the astronomers but did not bother to read the articles of the British paleontologist Alan Charig and his colleagues who wrote for various scientific journals, meeting the charge of their opponents. The technical questions raised by the astronomers were easily answered by five eminent paleontologists who wrote a paper entitled "*Archaeopteryx* Is Not a Forgery." This effort was sponsored by the British Museum that felt it was necessary to respond to the accusations and especially since it was still sensitive after the Piltdown forgery - a faked human skull. In fact, many started to call *Archaeopteryx* the "Piltdown Bird."

The charge that Hoyle had made was more than simply an attack on one of the key "proofs" of evolution. It was also an assault on perfectly beautiful examples of natural fossils that should be studied in an empirical way to determine their significance to the cause of factual science. **As a creationist it bothers me that because of the results of this charge against the validity of the fossils, these specimens were locked away in the basement of The British Museum of Natural History. Thus, the fossils were not for public viewing nor were they open for research. The entire incident, I believe, places creationists in a bad light. They are further deemed as individuals fearful of the inconvenience that the fossils might have proved against their theory of creationism.**

Even creationists (I am one of them) would have to agree, after reading the material of the astronomers and paleontologists, the forgery charge was unfounded. This is why **Jonathan Wells** could write, "Although the significance of *Archaeopteryx* for bird evolution remains controversial, all parties to the current controversy agrees that the fossils are genuine" (*Icons of Evolution Science or Myth,* P.114). **Wells writes against evolution but agrees that the fossils are authentic.** In other words, whether you are from the creationist camp or the evolutionist camp, you should agree that the fossils are genuine. Not withstanding, for those readers not familiar with the controversy as regards to the question of the fossils being authentic or fake, the following information is for you.

Every creationist respects what the Center for Scientific Creation has done for the cause of creationism. I have heard the lectures of many speakers from this Center and their books align my library shelves. I support their cause and believe in the creationistic doctrine these scholars teach but every once in a while, I am not in agreement with what the Center advocates. For example, **the FAQs entitled "What Was Archaeopteryx?" I wish had never been put into circulation and for the following reasons: I do not believe it is necessary to support every writer simply because he or she is a creationist.**

Even such an eminent scholar as Fred Hoyle could be incorrect in his assumptions and especially when he writes about a subject foreign to his particular field. I know, for a fact, that certain issues are not carefully investigated by representatives of Creationism and that some of our scholars are just as guilty as evolutionists in presenting that side of the issue which best represents their biases and predispositions. After all, this is a natural part of the human makeup which tends to indict every person as being guilty when holding to a personal and particular point of view, while at the same time failing to view the same indictment as applying to his/her own person.

Also, I believe that both sides of an issue should be investigated. In the case of the *Archaeopteryx* fossils, the book of Hoyle and Wickramasinghe on *Archaeopteryx* forgery was read by the Center but with few exceptions, even reflecting upon the opposing arguments. I come to this conclusion when considering the fact that *most creationists* are satisfied with the reasons set forth by the evolutionary paleontologists why they believe the fossils to be valid. I write this only to advance the truth of the *Archaeopteryx* fossils; not to further the cause of evolution.

I do not believe (as a creationist) that the *Archaeopteryx* fossils hurt the cause of creationism. The fossils are interpreted by evolutionists and set forth before the public eye to advance the cause of blood relationship between classes of animals and to promote the cause of transmutation of species. This can hurt the people's understanding of creationism because the meaning of the fossils has been misconstrued. But it isn't the fossils themselves that hurt the cause of creationism. There is nothing innate within the fossils that would make them ruinous to creationism's foundational beliefs, to make them inconvenient to the theory of creation, or to make creationists fear of not being able to respond to the geological, biological, and paleotological interpretations set forth by evolutionists. **The very purpose for my writing this book is to give people an understanding that the *Archaeopteryx* fossils can be accepted as genuine while at the same time, advancing the arguments for Creationism and invalidating the suppositions of Evolutionism.**

Creationists have nothing to be embarrassed about when it comes to the *Arxchaeopteryx* fossils. Our empirical approach to these fossils makes them an inconvenience only to the evolutionists who continue to interpret these specimens outside the facts of science. It is one thing to accept the validity of the fossils and view them in the light of the facts of nature; quiet another thing to advocate the evolutionary concepts that surrounds *Archaeopteryx* such as this creature is the missing link between reptiles and birds and modern birds are the direct descendants of small, advanced carnivorous dinosaurs, etc.

I will give a short rundown of the views held on the London and Berlin specimens of *Archaeopteryx*. Perhaps, I can help the readers to decide the scientific state of the fossils; whether they are natural or fake. In 1995-1997, the Center for Scientific Creation circulated FAQs titled "What Was *Archaeopteryx*?" The various statements of this work (creationism versus evolution) will not be presented in order to get to the main and salient points dealing with the fossils themselves. The best way to do this is to list the three major arguments advanced by Fred Hoyle and Chandra Wickramasinghe in their work "*Archaeopteryx, the Primordial Bird: A Case of Fossil Forgery.*" The arguments appearing in the FAQs are taken from this book. Apparently the Center and its main advocates considered Hoyle and his colleagues to have a strong case for the fabrication of the fossil bird.

The three major points of the argumentation will be given following this brief description of the chief allegation:

"Thin layers of cement were spread on two fossils of a chicken-size dinosaur called *Compsognathus*. Bird feathers were then imprinted into the cement."

FAQ, P.1 of 5 [That is, the feather impressions associated with the small skeleton were fabricated – Krishtalka's comment]

[Special Note: *Compsognathus* ("Pretty jaw") is one of the smallest known dinosaurs. It could reach up to 2 feet – more than half of this length being made up by a long, thin tail. A single skeleton was found in Germany in 1861]

The evidence of this so-called fossil quackery comes from the short trilogy of contentious observations now considered. **Following each observation there will be brackets containing comments that I have extracted mainly from the writings of Leonard Krishtalka and his book – _Dinosaur Plots and Other Intrigues in Natural History_. Asterisks will set off the various points of the FAQs taken from Hoyle's book.**

* "Many feather imprints show what has been called 'double strike' impressions. Apparently, feather impressions were made twice in a slightly displaced position as the slab and couterslab were pressed together." FAQ P.2 of 5 **[In other words, the "double-struck" appearance of the feather imprints are allegedly the result of a botched forging job rather than natural preservation]**

* "Evidence of a forgery includes instances where the supposedly mating faces of the fossil (the main slab and couterslab) do not mate."

FAQ P.2 of 5 **[That is, the forger had manipulated the fossil layers after the rock had been split open by the quarry workers. In the London specimen there is a poor fit of the main slab and counterslab]**

* "The feather impressions are primarily on the main slab, while the counterslab in several places has raised areas that have no corresponding indentation on the main slab. These raised areas, nicknamed chewing-gum blobs are made of the same fine grained material that is found only under the feather impressions. The rest of the fossil is composed of a courser grained limestone." FAQ P.2 of 5 **[Fabrication is the conclusion drawn from the contrast of the finer-grained nature of the sediment bearing the feather impressions and the coarser sediment embedding the bones]**

The three major points stated above are the purported evidences for forgery of the _Archaeopteryx_ fossils. Since these are the claims invented by Sir Fred Hoyle, the eminent British astronomer, many opponents feel that he should have stuck to star-gazing rather than step over into the field of paleontology of which he knows precious little.

A response to the aforementioned signs of fossil tampering will now be made by the noted paleontologist Leonard Krishtalka, an expert in his field. He is one of many who could take up the simple task of clarification. **Krishtalka claims that the arguments presented by Hoyle _can be explained by freshman paleontology_. Rather than abridge those remarks of Krishtalka and take a chance on** watering down the force of his rejoinders, Hoyle's arguments will be refuted by Krishtalka's own words.

Krishtalka speaks to the "three telltale signs of fossil shenanigans" invented by Hoyle and Wickramasinghe. **The signs of tampering are placed in italics:**

1. *"Double-struck" appearance of the feather imprints* – "Only Hoyle and colleagues are dumbstruck by the 'double-struck' feather impressions. As de Beer and others have explained, *Archaeopteryx* had *two* rows of slightly overlapping feathers on each wing, which would be preserved as overlapping natural impressions and appear double-struck. Also, in a quirk of preservation, the feather impressions on both the main slab and counterslab of the London specimen are the *underside* of the wing. A logical forger would have faked one of the slabs with the upper side of the wing and the counterslab with the underside.

"Moreover, Hoyle and associates never explain how the forger managed to produce natural casts of the feather impressions on one slab and identical negative casts at exactly corresponding position on the counterslab."

Dinosaurs Plots and Other Intrigues in Natural History

P.99

2. *"Poor fit of slab and counterslab"* – "The poor fit of the main and counterslab of the London *Archaeopteryx* is the result of scientific, not surreptitious, [and] tampering. Ever since the British Museum acquired the specimen, continued preparation of the slab (to expose more bones and feather impressions) and casting (to produce replicas for other museums) have altered the matching contours of the fossil-bearing surfaces.

"It's also obvious that Hoyle, Wickramasinghe and friends haven't done much rock splitting. When stone slabs are cracked apart they fall from geologic grace: Bits of loose rock crumble away and, like Humpty Dumpty, the slabs can't be put back together again quite as perfectly."

[Ibid. P.100]

3. *"Finer-grained nature of the sediment bearing the feather impressions in comparison to the coarser sediment embedding the bones."* – "The limestone around the feather impressions is indeed finer than that around the bones of the skeleton, but this comes as no surprise to geologists and paleontologists. The Solnhofen limestone, like most water-lain deposits, is composed of both fine- and coarser-grained layers. Fine structures, especially feather impressions, demand burial in fine-grained sediments to be preserved. If the grains composing the rock aren't fine enough, neither are the fossils. Also, algae, bacteria and fungi often coat the feathers and fur of decaying animal carcasses and trap much finer sediment than does the skeleton."

[Ibid. P.100]

Although these responses may make a person wish that he had studied freshman paleontology, nevertheless, they are clear and seem to be more than adequate for upholding the veracity of the original naturalness of the *Archaeopteryx* fossils and before they became man-handled in the production of replicas

There are other facts mentioned by Krishtalka which have a strong bearing on the issue and should put Hoyle's charges to rest.

Three of the facts can be summarized as follows:

1) In 1955, another partial skeleton of *Archaeopteryx* with faint but definite feather impressions was excavated near the site of the London specimen **(Were these feathers also forged?).**

2) In 1973, a fossil in Eichstatt's Jura Museum was reidentified as *Archaeopteryx*. The paleontologist F.X. Mayr noticed weak outlines of wing and tail feathers (Again, were these feathers also forged?).

"The obvious question is, did the forgers of 1861 and 1877 lay a plot that passed down through five generations and hatched ninety years later?"

[Ibid. Pp.100-101]

[The 1877 date refers to the Berlin fossil of *Archaeopteryx* which was discovered sixteen years after the London fossil. It was a more complete skeleton and found in another Bavarian quarry about ten miles form the first specimen. Attached to the outspread wings were impressions of primary and secondary flight feathers. Each vertebra of the long tail bore a pair of tail feathers]

David A.E.Spalding writes in *Dinosaur Hunters*:

"The accusers (Hoyle and his colleagues) produced photographs of the London specimen that, they suggested, showed blurring of the feather impressions, a gap between slab and counter slab, and different grain size for the sediments surrounding bones and feathers. This evidence was alleged to show that the fossil was a fake. As the Haberleins (father and son) had been involved with both fossils, it was easy to suggest that they were perhaps responsible, and it was hypothesized that the unpopular anti-evolutionist Owen knew about and perhaps even commissioned the forgery."

P.42

Spalding goes on to write that the technical questions raised by the accusers were easily answered and then brings up an issue much like the ones Krishtalka brought up above in the form of questions.

".... and they failed to explain the discovery of new specimens, long after Owen and the Haberleins were dead, or the rediscovery of old specimens collected before *The Origin of Species* was published. Although Owen was unpopular with many of his contemporaries, there is no evidence that he stooped to forgery to prove his anti-evolutionary agenda."

[Ibid. 42]

The explanations for feather impressions found in new discoveries or in the rediscovery of old specimens eluded the accusers and thus, there were no rejoinders that could be offered. I suppose it is easier to charge men who are resting in their graves and unable to meet the accusations than it is to face men openly and have to experience the force of their answers that would have turned seemingly good rationalizations (according to the accusers) into the banalities that they are.

3) Finally, Krishtalka points out how unreasonable it is to conclude the London and Berlin *Archaeopteryx* fossils were fakes:

"For that matter, if Hoyle is right, the fossil forgers must have been anatomical geniuses. Solnhofen yielded skeletons of small dinosaurs and *Archaeopteryx* that are virtually identical. Yet the forgers managed to doctor only those few small, dinosaurlike skeletons *with wishbones* and (except for the Eichstatt specimen) avoided feathering the more common dinosaurlike skeletons *without wishbones*. There are expert paleontologists practicing today that aren't that good."

[Ibid. P.101, Emphasis his]

I respect the professional level of Sir Fred Hoyle and Professor Chandra Wickramasinghe. Both scientists are eminent in their own chosen area of science but it is an unfortunate circumstance that these two wandering stars left their heavenly studies of astronomy to dabble with paleontology of the earthly realm. To demonstrate that this act was inopportune take the following example! After these two astronomers examined one of the *Archaeopteryx* fossils from southern Germany, they concluded that it was

34

fake (as we well know) and really the skeleton of a genuine dinosaur called *Compsognathus* with chicken feathers arranged around it. A limestone mixture supposedly had chicken feathers pressed into it before the mixture had set. This had accounted for the feather impressions. Dr. David Norman is Head of Paleontology at the Nature Conservancy Council of Great Britain and an Honorary Research Fellow at the University Museum, Oxford. He adds his opinion with reference to the so-called fossil forgeries and Professor Sir Fred Hoyle and Professor Chandra Wickramasinghe:

"Both these scientists ...have not taken into account the weight of geological and paleotological evidence in existence. As has been demonstrated by a number of studies, *Archaeopteryx* is nothing like *Compsognathus* in detailed structure and it is foolish to believe that paleontologists would be confused over this point. As for the faking of the feather impressions, it is perhaps naïve to suggest that scientists cannot recognize a fake when they see one, especially one where the features (*i.e.* the feathers) are of crucial importance.

"They believe ... that dinosaurs became extinct in a viral epidemic which came from outer space. The genes from such viruses grafted themselves on to animals which survived the plague at the end of the Cretaceous, transforming them into birds."

The Prehistoric World of the Dinosaur

David Norman

P.177

It appears that Hoyle incorporated some of his science fiction novels into events that occurred at the end of the Cretaceous Period. Even most creationists would find Hoyle and Wickramasinghe to have fostered ludicrous and ridiculous conclusions about *Archaeopteryx* and other birds. I have faith in my readers to conclude along with me that the *Archaeopteryx* fossils are genuine. Based on this assumption, we can move on to further studies of this fascinating and intriguing creature. The truth of the validity of the London and Berlin *Archaeopteryx* specimens should be *accepted* – We have already done this; the so-called "truth" of their evolutionary implications should be *challenged* – We have yet to do this.

Readers may have an interest in the subsequent prepared statement formulated by evolutionists:

"We – the present official custodians, preparatory and photographer of the holotype – reject this forgery hypothesis unequivocally. It may seem that we, in refuting the consortium's allegations, are using a sledgehammer to crack a rather trivial nut; yet, if we bear in mind the high esteem in which the general public holds Professor Hoyle, together with its lack of knowledge of the facts concerning *Archaeopteryx*, then it is important that such doubts be finally removed – especially where students of zoology are concerned. More important still, we must put the record straight because of the Creationists, who are interested in any new ideas that, implicitly or explicitly, appear to threaten the concept of organic evolution."

Charig, Alan J., Frank Greenaway, Angela C. Milner, Cyril A. Walker, and Peter J. Whybrow, "Archaeopteryx Is Not a Forgery," *Science*, vol.232 (May 2,1986), pp.622-626.

[Quote is found in the book "That Their Words May Be Used against Them" by Henry M. Morris, P. 623]

I have read this statement carefully and have concluded that it is as fair as possible. But they couldn't resist the metaphor, "a sledgehammer to crack a rather trivial nut." In this case, who can blame them? The case of fossil forgery was rather weak to begin with and was refuted with "Freshman Paleontology 101." However, I am not impressed with their concern for students of zoology; not when I known evolutionists are inclined to teach with "shocking dishonesty" the fraudulent theory of recapitulation; the transformation of ape to man when there isn't one thread of empirical evidence throughout the whole field of paleoanthropology; the ridiculous evolutionary belief that a hairy land animal could make the sea his home and eventually turn into a whale; the old and antiquated illustration of the "horse series" as indicating progressive transformations; the Miller-Urey experiment as demonstrating the first step in the origin of life; and the classic story of natural selection as illustrated through the peppered moth landing on light- and dark-colored tree trunks, allegedly proving the mechanics of evolution. **All of these theories are filled with flaws and yet evolutionists continue to flaunt them and much more before these same "students of zoology." I just wanted to "put the record straight." But, nevertheless, I go along with the evolutionists and their defending the validity of the *Archaeopteryx* fossils.**

There are other creationists who agree with me. For example Hank Hanegraaff who has earned the respect of all creationists. He is an

impressive crusader against the sophistry of Darwinian Evolution. He gives advice to creationists on how **not** to defend the Christian faith. He calls certain moves in the field of apologetics "death moves." These are moves the Christian should avoid. As a case in point he gives the illustration of a student talking with an evolutionary biologist who just cited the **Archaeopteryx** as a transitional form. Hanegraaff writes:

"Defensively you blurt out, '*Archaeopteryx* is really just a hoax.' Unfortunately, at this point your witness has been compromised. While *Archaeopteryx* is clearly not a transition from reptiles to birds, it is just as clearly not a hoax. Rather than speak out of ignorance, it is far better to admit that you haven't researched *Archaeopteryx* but you will, and return with a reasoned argument."

The Farce of Evolution

Hank Hanegraaff

Pp.126-127

Hank Hanegraaff has accepted the validity of the ancient bird fossils but continues to fight against their evolutionary implications. I trust my readers will do the same after they gain more information.

I am fearful that many creationists will feel that I have taken sides against the astronomers Sir Fred Hoyle and Professor Chandra Wickramasinghe. In the general battle for the truth of *Archaeopteryx* being a fake, I feel that the astronomers have lost. In the specific war for the truth of creationism, I feel that the paleontologists have lost. I am very thankful that Hoyle and Wickramasinghe, who were once atheists, now accept creationism and the God of creation. Almost twenty years ago these two astronomers, who were friends, became interested in life's origin. They made calculations concerning the chance of life developing throughout the universe and apart from God. They concluded the probability of life evolving anywhere in the entire cosmos was the number 1 followed by 40,000 zeroes. In other words, there was no probability. James Perloff has written a fascinating book entitled *Tornado in a Junkyard*. This title was taken from the past words of Sir Fred Hoyle when he said, "the probability of evolution is equal to the probability that a tornado, sweeping through a junkyard, would assemble a Boeing 747!" [The fact that parts for a Boeing 747 would not be found in a common junkyard (Isaak); does not invalidate the force of Hoyle's argument]

When the two professional astronomers stepped across the line and turned their backs on evolution, they must have had a profound influence upon their colleagues as well as the public. Hoyle and Wickramasinghe manifested courage and daring – traits of character found in the lesser number of professionals of the scientific profession. These men followed not only their conscience but logic and common sense in their assessment that whenever there is life in the universe, it had to be created (See Romans 1:20). Thus, Intelligent Design *might account* for the complex things in life's creation but *only* GOD, in a positive and sure manner, could bring forth the complexity of life itself.

[Pittack's Note: Mark Isaak, in his book *The Counter-creationism Handbook* devotes sixteen pages to a refutation of Intelligent Design. He claims on page 250 that Intelligent Design implies results that are contrary to common sense. He cites spider webs as his main focus in making creationists and their theory of intelligent design appear to be ludicrous and inane. Since, due to time limits, I can not rebut all sections of his work, I would care to respond to his charge concerning cobwebs since spiders have always been sources of fascination for me. The evolutionist all the time raises the question of who created or designed God. They reason if, as some claim, God does not need a cause, then by the same reason, neither does the universe or things that appear to be designed. This "prehistoric" argument will be discussed in Appendix II and III]

CHAPTER 3

ARCHAEOPTERYX – THE "PERFECT" LINK BETWEEN REPTILES AND BIRDS

(Is *Archaeopteryx* Truly the Missing Link?)

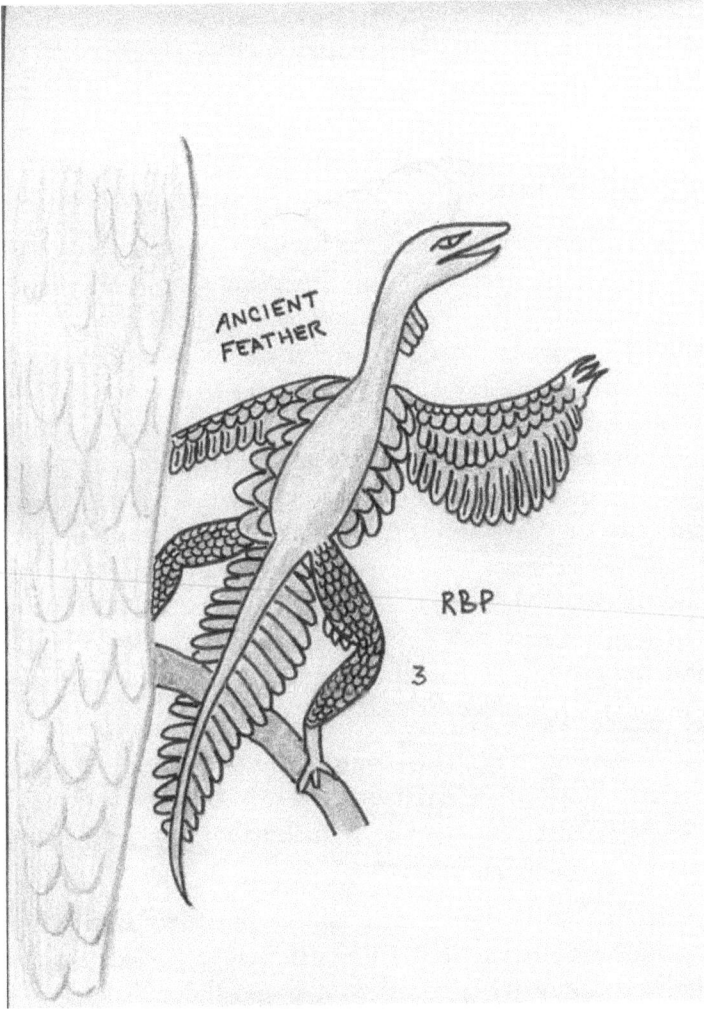

"And to all the *beasts of the earth* and all the *birds of the air* and all the creatures that move on the ground – everything that has the breath of life in it – I give every green plant for food. And it was so."

Genesis 1:30

NIV

Archaeopteryx will now be investigated to determine the real particulars with reference to its evolutionary status. Robert Bakker's remark that "Creationists…never allow their faith to fall victim to facts" is the same thing as charging that creationists persist in their belief in creationism no matter how strong the case is for evolution. Bakker would lead the public into thinking that creationists are stubborn and headstrong but evolutionists are deep thinkers with a well stocked arsenal to back them up. In point of fact when it comes to *Archaeopteryx*, it is one of the last fossil hopes of evolutionists for a connecting link between reptiles and birds. In reality their arsenal is extremely ill-stocked when it comes to connecting links between any classes of animals in the fossil record.

"The Counter-Creationism Handbook" written by Mark Isaak and pages 113-116 gives examples of fossil transitions between species and genera and even between families, orders, and classes. He also cites examples of transitionals between kingdoms and phyla. I was not impressed by Isaak's examples and, in this book, I will pay strict attention to one of his alleged transitionals … the dinosaur-bird transition. He claims that his examples demonstrate evolution: one type of animal changing onto another type of animal but special creationists knows that separate kinds of animals remain separate. Variation is not evolution and can do nothing more that erect additional races and variety within the created kinds. A "new species" is not a "new kind" of animal but merely a variation within the created kind. Again, for emphasis, variation *cannot* occur in unlimited fashion and create new kinds of animals.

In the middle 1980's a Conference was held in Germany to determine the status of *Archaeopteryx*. This conference was held at Eichstatt, Bavaria, September 11 to 15 and included research reports from most of the world's specialists on the earliest bird. They were united as they expressed the following doctrine:

"Archaeopteryx was a bird that could fly, but it was not necessarily the direct ancestor to modern birds. It was a bipedal cursor that was facultatively aboreal. Flight developed with the assistance of gravity (e.g., from the trees down) rather than against gravity (from the ground up). *Archaeopteryx* was probably derived from theropods. A communiqué expressed the unanimous belief of all participants in the evolutionary origin and significance of *Archaeopteryx* was adopted, in order to forestall possible misuse by creationists of apparent discord among scientists."

[Dodson, Peter, "International *Archaeopteryx* Conference," *Journal of Vertebrate Paleontology*, vol. 5 (June 1985), pp. 177-179]

[Special Note: The members were weary of creationists constantly charging them with not being in harmony on their evaluation of *Archaeopteryx!* Therefore, they banded together and developed a statement to the effect they collectively believed in the evolutionary origin of *Archaeopteryx* and its significance as an intermediate between reptiles and birds. This communiqué was nothing new for the field of evolutionary science. When evolutionists spend their valuable time at a conference dedicated to *Archaeopteryx* and they do not come up with anything that hasn't been said over the last century and a half, the relevancy of their theory appears to be in trouble]

Forgive me for mentioning the fact that creationists not only take advantage of the evolutionists' plight of their not being in harmony with their beliefs concerning *Archaeopteryx* but also take advantage when they profess that they *are* in harmony. For example, the specifics drawn up in this credo of origins and characteristics of the ancient bird are highly contested at this present time. There are scientists who believe this bird was unable to fly; many of those who do believe in his flight ability do not agree that the origin of flight for the *Archaeopteryx* began from the trees down; who do not believe this bird to be derived from small theropods. **No signed creed will ever stand in place of investigative science and serve as an assurance to every scientist that he is in pure harmony with his colleagues on the facts of nature pertaining to Archaeopteryx.**

On the one hand, it is erroneous to think that scientists who come up with the same facts, automatically places them at the goal of empirical science. On the other hand, science should not deter any scientist from having conflicting thoughts with fellow scientists. This is the way of true science – to have men come up with various ideas and then put these ideas to the test of experimental investigation. **What earthly good is it to the cause of science to have men sign in agreement that they share the same views on some fossil that is supposed to have lived 150 million years ago? Are the creationists of such a dim-witted frame of mind that they fail to see the loopholes in all this so-called harmony bit?**

Are evolutionists, in truth, able to advance the notion that biological characteristics can be properly assessed for fossils millions of years in age? Is the mind thinking along evolutionary principles automatically capable of establishing omnipotence when it comes to human knowledge? The field of paleontology is filled with all types of theories concerning *Archaeopteryx*. This book will seek to mention some of them and especially how they are viewed under the spotlight of creationism. Although the thinking can be somewhat repetitious and cumbersome, a number of important evolutionists will be cited for their conclusion regarding *Archaeopteryx* as the link between reptiles and birds. This is the most important feature to be considered since it is nearly a universal claim proposed among evolutionists.

Over one hundred years ago, Sir John Evans (1823-1908) was an evolutionist who would not be deterred or discouraged by the initial account given by Richard Owens of *Archaeopteryx* being *only* the first bird. In 1865, Evans wrote that the fossil bird was extremely important because of its paleotological significance. It attested to the great question of the origin of species and "seemed to link together the two great classes of Bird and Reptiles."

[This is no different from the 1985 communiqué mentioned above and drafted 120 years later - *Archaeopteryx* originated by progressive evolution and it is a link between birds and reptiles]

Richard Owens was a creationist and he is often criticized in evolutionary literature as not making a thorough investigation of *Archaeopteryx*. But this charge is unfounded. First of all he was not new to fossil birds; he had already studied the giant Moas from New Zealand. He was England's greatest anatomists and known for his meticulous research and detailed examination of fossils. Secondly, when he reported to the Royal Society that *Archaeopteryx* was unequivocally a bird, though one of a primitive kind, he reported only what present day scientists conclude today. Owens is criticized for reporting the fossil was a bird ONLY but Ostrom and Swinton, who are modern-day bird experts, voice the same opinion. Ostrom wrote, "There can be no doubt that *Archaeopteryx* was a true bird" and Swinton says, "There is no fossil evidence of the stages through which the remarkable change from reptile to bird was achieved." [Both of these quotes are found in "Darwin's Leap of Faith" by John Ankerberg &John Weldon, Pp. 221-222]

In 1868, Thomas Huxley produced an article which examined the similarity between the small dinosaur *Compsognathus* and the "dinosaur bird" *Archaeopteryx*. Huxley also lectured on evolution and cited the *Archaeopteryx* as a missing link. *Compsognathus* was a small, bird-like dinosaur that looked a bit like *Archaeopteryx*. Huxley regarded this small dinosaur to be an important link between birds and reptiles.

[Special Note: Huxley considered *Archaeopteryx* to be an important discovery in offering evidence for Darwin's theory but he considered

Compsognathus to be even a closer estimate to the "missing link" and even suggested that birds evolved from dinosaurs. Modern studies have rejected *Compsognathus* as an ancestor of birds because it was the same age as *Archaeopteryx*. The study of modern cladistics brought in more problems as many "paleontologists found that the most likely candidates for the ancestor of *Archaeopteryx* lived tens of millions of years *later*" (*Icons of Evolution* by Jonathan Wells, P.121). Another reason for rejecting *Compsognathus* as the ancestor of *Archaeopteryx*; there are too many dissimilarities. In chapter two, Dr. David Norman said, "....*Archaeopteryx* is nothing like *Compsognathus* in detailed structure...."

Othneil Charles Marsh of Yale, the *American* counterpart of Huxley, also recognized *Archaeopteryx* as a missing link between two widely separated groups. He wrote:

"The classes of Birds and Reptiles, as now living, are separated by a gulf so profound that a few years since it was cited by the opponents of evolution as the most important break in the animal series and one which the doctrine could not bridge over. Since then ... this gap has been virtually filled ... *Compsognathus* and *Archaeopteryx* ... are the stepping stones by which the evolutionist of to-day leads the doubting brother across the shallow remnant of the gulf, once thought impassible."

[Othneil's quote is in *Dinosaur Hunters* by David A.E. Spalding, P.35]

Robert Bakker has already been cited. Here, again, is his quote concerning *Archaeopteryx*:

"Inveterate creationists, then or now, never allow their faith to fall victim to facts. But to any careful, unbiased observer, it was clear that the fossil bird from the Age of Reptiles consisted of a genuine link between classes."

The Dinosaur Heresies

P.302

Bakker views the fossil bird of 1861 as a "missing link" between the two distinct classes of reptiles and birds. He believes the anatomical structures of *Archaeopteryx* and its position in the geological strata meet the requirements for establishing the reptile-bird as a genuine intermediate.

Bakker is so carried away and elated over his facetious remarks concerning creationists; he fails to mention essential matters. With all his fuss over *Archaeopteryx* being an intermediate link between reptiles and birds, he fails to talk about its being, in past

life, a fully functional organism. This "stratomorphic intermediate" does not possess intermediate structures. *Archaeopteryx* may appear to be a paleotological link but his feathers, teeth, and claws are all completely functional.

Bakker fails to tell us how this "intermediate" fossil bird containing *no intermediate structures*, can be used as evidence to advance his evolutionary scheme. *Archaeopteryx* was a fully functional organism with fully functional structures. In other words, it is not an intermediate between reptiles and birds, not to be perceived as a transitional fossil between two classes of animals, not to be waved before the creationist as the total proof of a connecting link that would deem evolution as a foregone conclusion. Ernest Lutz claims:

"There is neither evidence of lineage from reptiles to *Archaeopteryx* nor it to any living birds … In view of the evidence, science has oversold the case for *Archaeopteryx* as a transitional form."

["A Review of Claims about *Archaeopteryx* in the Light of the Evidence,"

Creation Research Society Quarterly, June 1995, P.18]

In reply to Robert Bakker's facetious remark mentioned above, the words of those two researchers (John Ankerberg & John Weldon) come to mind – "*Archaeopteryx* as a transition just doesn't fly" (*Darwin's Leap of Faith*, p.221).

Leonard Krishtalka, the paleontologist who argued for the validity of the *Archaeopteryx* fossils, has this to say about them:

"They are not freaks of nature. They are thumping confirmation of an evolutionary continuity between 'kinds' …

"*Archaeopteryx* was the perfect 'missing Link' between birds and reptiles and dramatic fossil evidence for the continuous evolutionary thread between different kinds of animals."

Dinosaur Plots and Other Intrigues in Natural History

Leonard Krishtalka

Pp. 102, 96

Like Bakker, Krishtalka understands *Archaeopteryx* to be not only the "missing link" between reptiles and birds but the "perfect" link. **Krishtalka has given a satisfactory investigation of the ancient bird fossils to indicate they were not fraudulent and we have accepted his first-rate analysis but when he writes of *Archaeopteryx* being** the "perfect" missing link or its being "evidence for the continuous evolutionary thread between different kinds of animals," the creationist must part company with Krishtalka's way of thinking.

Krishtalka has left empirical investigation and has entered the realm of evolutionary philosophy. These statements from Krishtalka sounds more like wishful thinking on his part rather than observation based on factual evidence. In the first place the ancient bird fossils are not testimonials of the missing link (this book will soon attest to the reasons why they are not). And in the second place how one example of a link can be cited as "evidence for the continuous evolutionary thread between different kinds of animals" is an incongruous statement to say the least.

If anything has been evidenced in the field of paleontology, it is the undeniable fact that there are no transitions between classes of animals. To use *Archaeopteryx* as the *Exemplar* of a "thread" *linking all classes of animals together is unscientific.* Krishtalka, by this one example, seeks to support two main tenants of evolution such as: beings now living have descended from different beings which have lived in the past and the gaps now existing between clusters of forms have arisen over millions of years and if we were able to view time speeded up, these same gaps would be non-existent. That is, if we were able to assemble all the life forms which have inhabited the earth and view them together, they would be seen as a continuous arrangement without gaps or breaks (Marsh). This, of course, could never be done and that is why evolution must always remain a philosophy and without empirical evidence to back it up.

Therefore, the evolutionists who accept gradualism or gradual transitions between classes of animals are not able to say, with a clear academic conscience, they derive this information from the fossil record. There are no verifiable transitions from one form of life to another. This phenomenon of evolutionary change that supposedly forms one kind of creature from a completely and totally different type of creature can be found neither in living species nor in fossil remains. The following words from a couple of famous evolutionists should confirm the reality of this fact. Hundreds of like admissions could be gathered from other evolutionists but writings from these two, should suffice.

A.J. Boucot writes:

"Since 1859 one of the most vexing properties of the fossil record has been its obvious imperfection. For the evolutionist this imperfection is most frustrating as it precludes any real possibility for mapping out the path of organic evolution owing to an infinity of 'missing links'… once above the family level it becomes very difficult in most instances to find any solid paleotological evidence for morphological intergrades between one suprafamilial taxon and another. This lack has been taken advantage of classically by the opponents of organic evolution as a major defect of the theory…. the inability of the fossil record to produce the 'missing links' has been taken as solid evidence for disbelieving the theory"

[*Evolution and Extinction Rate Controls* (Amsterdam: Elsevier Scientific Publishing Co., 1975), 427 pp. Quote of Boucot is found in *"That Their Words May Be Used Against Them* by Henry Morris, P.161]

David B. Kitts writes: "Few paleontologists have, I think ever supposed that fossils, by themselves, provide grounds for the conclusion that evolution has occurred. An examination of the work of paleontologists who have been partially concerned with the relationship between paleontology and evolutionary theory, for example that of G.G. Simpson and S.J. Gould, reveals a mindfulness of the fact that the record of evolution, like any other historical record, must be construed within a complex of particular and general preconceptions not the least of which is the hypothesis that evolution has occurred."

["Search for the Holy Transformation," review of Evolution of Living Organisms, by Pierre-P. Grasse, *Paleobiology*, vol.5 (Summer 1979), pp.353-355. Kitts was Professor of History of science, University of Oklahoma. Quote of Kitts is found in *"That Their Words May Be Used Against Them"* by Henry Morris, P. 193]

The fossil record has produced nothing that can be identified as a missing link. Krishtalka cites *Archaeopteryx* as the possible *Exemplar* of the thread that links all classes of animals in the fossil record. But it can not even serve as the thread connecting reptiles and birds which we will soon discover. For evolutionists to appeal to this one example when countless thousands (at least a quarter of a million) of missing links are needed to fill in the gaps of the geological column, demonstrates the futility of their doctrine.

The only way that one could believe that major phyla cross over the geological lines of such major discontinuities would be for one to believe in evolutionary miracles. No small wonder that the fossil record which provides such a sparse amount of information on transitional forms has become a great embarrassment for those who believe in organic evolution. Gould says: "The extreme rarity of transitional forms in the fossil record persists as the trade secret of paleontology" (S.J.Gould, 1980. Quote in *Paleobiology* 6, 119-130).

However, it is no longer a trade secret. Biologists, along with other professional scientists, have been alerted to the idea the fossils do not provide evidence of gradual evolutionary change.

Not only that, but the *Archaeopteryx* has been "dethroned" (according to Jonathan Wells). The fossil bird has been removed from its iconic status as a missing link.

But before learning about the "dethronement" we must speak to the issue of punctuated equilibrium. Since paleontologists come up empty handed when appealing to the fossil record for support of transitional forms of life; why don't all of them accept the doctrine of quantum leaps from one species to another! At the very least, they would be enabled to explain (?) the gaps or stasis when species exhibit no directional change and the sudden emergence of classes that appear all at once and fully formed. **Think about the reasons why evolutionists do not accept the theory of punctuated equilibrium! There are at least four reasons:**

1) In the first place it is hard for them to turn from the traditional time it takes to evolve an entirely new form of life. Gould and Eldredge picture evolution as bursting all over the earth in different times and places. These bursts would have a lengthy hiatus or a long period of stasis. However, these new forms (new species) would appear over thousand of years rather than hundreds of thousands or millions of years. This "time" would not be "deep" enough in the thinking of most evolutionists, for the transmutation of new species. There wouldn't be enough time since the evolutionary process is a very slow one involving chance mutations and the slow work of the mechanical power of natural selection.

Some creationists have claimed that accepting punctuated equilibrium would be like accepting the "hopeful monsters" theory purported by the geneticist Richard Goldschmidt. For example, the first bird hatching from a reptile egg. Many evolutionists have faith coupled with science but few have nerve enough to accept what they believe to be science fiction. Gould contradicts himself by stating that transitional forms are generally lacking at species level but are abundant between large groups and then reversing his thoughts by affirming at the higher level that gradualism has always been in trouble but not at the lower level. (See Gould's contradictory statements quoted in *Creation Scientists Answer Their Critics* by Duane T. Gish, Pp.135-136).

Another contradiction in Gould's way of thinking is found in his remarks of the "hopeful monsters" theory. In one instant he says, "Any evolutionist who believes such nonsense would rightly be laughed off the intellectual stage…." And in another he seems to support Goldschmidt's theory to the utmost. Gould labels "the first bird hatched from a reptilian egg" as nothing more than a simple metaphor used by Goldschmidt but Gish accuses Gould as backpedaling (For full coverage of this issue, please refer to Gish's above mentioned book, Pp.135-146).

2) In the second place evolutionists find it difficult to turn away from the concept of vertical transitions in the geologic column. The theory of punctuated equilibrium professes to be scientific but then explains why the evidence for it cannot be found. "The theory does not remove the need for transitional fossils. It only tells us why those transitions were never found" (*Darwin's Leap of Faith* by John Ankerberg and John Weldon, P.224).

3) In the third place, evolutionists find it difficult to turn from the science of genetics. Gish claims:

"This notion of punctuated equilibrium, which is being hailed by many as the solution to the problem posed by the fossil record, is actually no solution at all. First of all, punctuated equilibrium is not a mechanism. No one knows how a species could rapidly evolve into a new species. In fact, this notion is contrary to our knowledge derived from the science of genetics. The genetic apparatus of a lizard, for example, is devoted 100% to producing another lizard. The idea that this indescribably complex finely tuned, highly integrated, amazingly stable genetic apparatus involving hundreds of thousands of interdependent genes could be drastically altered and rapidly reintegrated in such a way that the new organism not only survives but actually is an improvement over the preceding form is contrary to what we know about the apparatus and how it functions."

[Only Gish could write a sentence containing over sixty words and still be remarkably clear and lucid on the subject at hand – genetics]

Evolution; the Fossils Say NO!

Duane T. Gish

P.355

4) In the fourth and final place, Gish writes:

"Furthermore, this notion is without empirically observable scientific evidence. The only evidence for it is the absence of transitional forms.

According to the punctuationist, since obviously one form did not slowly and gradually evolve into another, then just as obvious it must have rapidly evolved into the new form."

[Ibid. P.355]

The bottom line is this – No matter if the evolutionist decides to go to the fossil record to *derive attestation* for organic evolution or whether he elects to turn to the theory of punctuated equilibrium in order to *explain* organic evolution, he will come away empty handed from either experience. Gish concludes his opposing case with this conclusion – "The rising popularity of the punctuated equilibrium notion of evolution is just another indication of the bankruptcy of evolution theory."

[Ibid. P.35]

Nevertheless, the theory of punctuated equilibrium accomplished some good in attracting newspaper reporters and alerting many scientists to the non-existence of transitional creatures in the fossil record. With widespread attention of the gaps in the earth's strata, many evolutionary scientists have been caused to rethink their position. Every year there are evolutionists who take their stand for creationism. Also, consider the reporters who give accounts of scientists and their findings. For example, they report to the public domain, "In the last decade, however, geologists have found rock layers of all divisions of the last 500 million years and no [This partial quote is taken from the British newspaper *The Guardian Weekly,* 26 November 1978, vol 119, no22, p1 and found in *Evolution: A Theory In Crisis* by Michael Denton]

People around the world are being warned of the short comings of evolution and have their attention drawn to the Creator of heaven and earth. It is as though the voice of the first angel, in Revelation the fourteenth chapter of the Bible, is sounding throughout every nation: "Fear God and give him glory, because the hour of his judgment has come, Worship him who made the heavens, the earth, the sea and the springs of water."

Revelation 14:7

NIV

We will return to a few more famous evolutionists who can be cited for their conclusion regarding *Archaeopteryx* as a link between reptiles and birds.

Roger Tony Peterson, Doctor of Science writes:

"Certainly this first true bird retains many of the characteristic features of its reptilian ancestry. Its jaws were armed with strong teeth; the bones of the head and jaw are reptilian in character, and the bird retained a long reptilian tail. But this creature alone was enough to designate ....*Archaeopteryx* the first known bird – the furthest back man has yet been able to reach into the ancestory of the 8,600 bird species we know today."

The World Atlas of Birds

P.10

Peterson understands *Archaeopteryx* to have reptilian and avian features. He accepts the theory that this "first known bird" was an intermediate between two distinct classes of animals.

[Special Note: In 1983 the iconic status of *Archaeopteryx* as the first bird was *challenged* by Sankar Chatterjee. He and his co-workers claimed to have discovered two crow-sized birds in Texas rock of deep time supposedly dating back 225 million years ago. This presents a problem for such scientists as Roger Peterson whose statement was mentioned above. If Chatterjee's fossil birds are 75 million years older than *Archaeopteryx*, then how could *Archaeopteryx* be the ancestral bird? By the same token, these same fossil birds would be as old as the first dinosaur. How then could anyone hold to the idea that dinosaurs are the ancestors of birds? Obviously, *Archaeopteryx* is not an intermediate between reptiles and birds. Also, these fossil birds according to the evolutionary dogma should be much more reptilian-like in form since they are 75 million years closer to reptiles. But they are, in fact, more bird-like than the *Archaeopteryx*]

Christopher McGowan writes:

"If the skeleton of *Archaeopteryx* had been like that of a modern bird I doubt that it would have attracted very much attention. But here was a feathered animal with an unmistakably reptilian skeleton. It had long been recognized that birds and reptiles were closely related because they share a number of features, including the possession of scales and the laying of shelled eggs, here was solid fossil evidence of the link between two major groups."

Dinosaurs, Spitfires, & Sea Dragons

P.13

[Special Note: There isn't the slightest hint from the fossil record that the origin of birds can be traced back to reptiles. If this were so, there would

50

be countless intermediates tracing their lineage and outlining their heritage. The likenesses cited are superficial compared to their contrasting characteristics. The evolution of birds is far more complex than the above statement implies. The avian lung and respiratory system is so much different than that of all other vertebrates including the reptiles. And there is the problem of feathers developing from scales which will be considered later in this book. One more thing that is worth mentioning – practically all of the reptilian characteristics supposedly detected by evolutionists in *Archaeopteryx* are, when closely scrutinized, actually avian. This will be made clear in due time]

Coming up is the final example of an evolutionary viewpoint on the reptile-bird connection. Gregory S. Paul is a free-lance dinosaurologist and one of the well-known dinosaur artists. He writes in a facetious style:

"It is a remarkable coincidence that this most famous of the protobirds was first discovered just two years after the publication of Darwin's *The Origin of Species*. Darwin's detractors had been using the lack of any known links between reptiles and bird to challenge his theory. Those deities of a fundamentalist sort must have been in a self-destructive frame of mind, because *Archaeopteryx* gave the aggressive Thomas Huxley just the ammunition he needed to blast away at those skeptical of evolution."

Predatory Dinosaurs of the World

P.353

For Paul, not only has *Archaeopteryx* been accepted as a known link between reptiles and birds but placed into Huxley's thought pool: an evolutionary weapon to eradicate the God of creation and demoralize the doubters of evolution. **Is it not strange how one seemingly connecting link in the whole of the paleotological record, which lacks countless thousands of intermediate forms, suddenly becomes the answer to the phenomenon of discontinuity and turns out to be such a formidable weapon in the arsenal of the evolutionist who is now filled with an emotional confidence and self-pride in accepting evolution's credo. Talk about being "in a destructive frame of mind!"**

Many other statements could be gleaned from the writings of evolutionists to substantiate their belief that *Archaeopteryx* is an important fossil demonstrating presumed evolution from reptile to bird but those cited above will suffice. Apparently for the evolutionist, *Archaeopteryx* stands as definite proof that reptiles evolved into birds. This reptile-bird, supposedly, indicates a major transitional fossil half-way through a grand evolutionary change. Darwinists say, "At last, here is a fossil half-way between a reptile and a bird. The creationist should sit up and take notice for we have a definite link in the fossil record."

Creationists have taken notice but they conclude there is not much to note. The *Archaeopteryx* claim to fame is neither impressive nor remarkable and for the simple reason that modern cladistic analysis has had an important end result for *Archaeopteryx*.

Jonathan Wells states that cladistic analysis "removes the 'First Bird' from its iconic status as a missing link, and turns it into just another feathered dinosaur…so if birds are descended from dinosaurs then birds are dinosaurs."

Icons of Evolution

P.122

Some evolutionists, even now, refer to the *Archaeopteryx* as being dethroned from the ruler-ship of missing link status and as just another dinosaur with feathers. Again, Wells will be quoted for his eye-opening, heart-searching, and mind-boggling conclusion about the status of *Archaeopteryx* as a Missing link:

"Isn't it ironic that *Archaeopteryx*, which more than any other fossil persuaded people of Darwin's theory in the first place, has been dethroned largely by cladists, who more than any other biologists have taken Darwin's theory to its logical extreme? The world's most beautiful fossil, the specimen Ernst Mayr called 'the almost perfect link between reptiles and birds,' has been quietly shelved, and the search for missing links continues as though *Archaeopteryx* had never been found."

[Ibid. P.135]

Cladistic analysis is the new approach for classification of animal species and is made up by a certain branch of evolutionists. Creationists have absolutely nothing to do with this *turbulent theory for type taxonomy*. The evolutionists who are seeking to establish this "tool" have stirred up bitter controversy with other evolutionists. Should it be a source of wonder to those who advocate the doctrine of evolution, why creationists are not impressed with *Archaeopteryx* being cited as the link connecting reptiles and birds! This is a controversy which needs to be settled among evolutionists. Meanwhile, there is no more proof for the *Archaeopteryx* being a missing link than there is for the creationist theory that the "reptilian-bird" was one of the created kinds which has become extinct.

Frank Lewis Marsh makes this statement:

"The *Archaeopteryx*, or Lizard-tailed Bird, is often cited by evolutionists as a connecting link between reptiles and birds. Evolutionists and creationists will agree that it has some of the characteristics of both reptile and bird. But does this fact make it necessarily a connecting link between the two? The duckbill is like a mammal and like a reptile. Is it a connecting link between them? The gnu or wildebeest is like an ox and like a horse. Is it a connecting link between them? This is merely another case of subjective evidence which can be explained from either point of view."

Evolution, Creation, and Science

P.275

There are transitional problems arising out of a concept which views *Archaeopteryx* as a missing link. A series of such transitional problems will now be addressed in the following two chapters. In chapter four, the difficulty of finding intermediate fossils in the geological record will be mentioned in a general way. We will gain insight as to the major complications in the field of paleontology when it comes to establishing the validity of evolution.

Chapter five will deal with the more specific issues. For examples, Biochemistry which allegedly traces blood relationships of fossil forms and Cladistic Analysis which supposedly traces the various branches of family trees and shows how fossils and life forms are linked to a common ancestor.

I trust that it is not too early to mention chapter twelve will actually speak to the problem of tracing fossils through the *entire* Geologic Time Scale. This chapter is entitled "The Place of the *Archaeopteryx* in the Geologic Time Scale." Robert Bakker, the famous dinosaurologist and artist, claims that *Archaeopteryx* is a perfect "missing link" that can be correctly followed in its climb up through the geological column. He also makes the bold assertion:

"The stratigraphic proof for a Darwinian origin of birds appeared incontrovertible – the rocks preserved the stages of development in the exact proper sequence through time. Any impartial observer might conclude that if God had really created birds, he must have been going out of his way to fool humanity into believing in evolution."

The Dinosaur Heresies

P.303

The entire aspect of Bakker's bold assertion will be fully investigated in chapter twelve. We will learn how worthless his claim really is but for now, let us lead up to chapter twelve (the final chapter in this book) by continuing to follow, in a logical and systematic manner, the unfolding of the *Arxchaeopteryx* story.

CHAPTER 4

THE GENERAL PROBLEM OF PALEONTOLOGY
AND FOSSIL TRANSITIONS

(Are the Creationists Unreasonable?)

"For since the creation of the world God's invisible qualities – his eternal power and divine nature – have been clearly seen, being understood from what has been made, so that men are without excuse .... Although they claimed to be wise, they became fools and exchanged the glory of the immortal God for images made to look like mortal man and birds and animals and reptiles."

Romans 1:20, 22

NIV

Creationists have often been criticized for their incessant demand that evolutionists demonstrate transitional fossils in the geological record. But why has this appeal been deemed so unfair and unreasonable? It is because the evidence that creationists ask for can not be produced by the evolutionists. They face the same dilemma that Darwin faced in 1860 when he lamented over the fact that the absence of numerous transitional links in the geological record was the most obvious and serious objection against his theory of evolution. According to Darwin's theory, every geological formation and every stratum should have been filled with intermediate links. This, however, was not the case. Darwin's theory of progressive modification was thwarted by the geological column – it simply negated his doctrine. Darwin, in spite of this, maintained his hope for the future; the imperfection of the geological record would eventually work itself out and the discovery of many links would fill those dreaded gaps that made his theory so unworkable.

The "truth" of Paleontology was to hopefully serve as a testimony to the doctrine of progressive modification preached by the Master of organic evolution and believed by every Apostle who took his philosophy to heart. But, according to the pages of geologic "scriptures," recorded in the deep time of every stratum, it was a fossil fact that the earlier members of any long-continued group of extinct animal forms were no more generalized in structure than the later ones. In other words, there is no finely-graduated organic chain in any class of animal structures that shows a transformation of one type of animal into another type. Numerous species suddenly come on the scene and in full perfection.

They remain substantially unchanged and then pass away in full perfection, appear to be planted at each geological stratum, come into view not according to the evolutionary belief that life structures developed from simple to more complex forms but like the creationist belief that major groups arose simultaneously as they came forth from the hand of God.

Only two years after the publication of *Origin of Species* in 1859, the finding of *Archaeopteryx* (the fossilized bird of the Jurassic Period) made clear to evolutionists that this fossil bird from the Age of Reptiles consisted of a genuine missing link between classes. For Darwin, the avian fossil not only appeared to be an intermediate between two distinct classes but made its entry at the correct sequence of time. Darwin, at last, had his stratigraphic proof for the origin of birds – just as he predicted. It would just be a matter of time until the fossils would verify all his other predictions. Every fossil level would demonstrate the required transitional forms required to fill in the gaps between the long acting processes of evolution – so Darwin thought.

However, this side of Darwin's time, billions of known fossils have been discovered and identified. There are still millions of specimens awaiting their identification and for the informative, dynamic dissertations for scientific journals that will be written and prepared by providential, professional paleontologists. Museums and Research Centers are overstocked with fossils but aside from *Archaeopteryx* and a handful of candidates there are few alleged, intermediate forms preserved in the rocks when there should be multitudes. The same general problem of transitional forms exists today. Let us listen to a number of professionals as they establish this claim to be a true fact of science. The reason this first statement has been selected, it mentions our favorite bird – the *Archaeopteryx*.

Louis Jacobs writes:

"How can it be that the fossil record of bird origins is so topsy-turvy? Where are the older, nonfeathered coelurosaur ancestors of birds? This is a bother for those, like me, who would prefer the fossil record to reflect the sequence, species by species, of the evolution of major groups of animals. Unfortunately that is an ideal that the fossil record has been unable to live up to in many cases. If it already did that, there would be less reason to search for fossils. The true ancestral species may never be found, but so what? We cannot even say for sure that the species ancestral to birds was fossilized. In this case, rocks of the appropriate age

– that is, immediately preceding *Archaeopteryx* – and representing the necessary depositional environment to preserve bird ancestors are rare. Nevertheless the hypothesis, based on the anatomy of birds, dinosaurs, and all other animals, remains that all birds, not just *Archaeopteryx*, are coelurosaurs, and that they evolved specifically from an older, as yet unknown ancestor. This provides a prediction, as good hypotheses do, that such an animal may be found…"

Quest for the African Dinosaurs

Pp.168-169

The main supposition of the above statement is birds are coelurosaurian dinosaurs. And how does one arrive at this brilliant deduction? Certainly, not by appealing to the empirical evidence of sequential arrangement of ancestral species; such an array has never been found in the fossil record. Jacobs infers: why bother with the paleotological record which seldom accommodates progressive theories? We must not look for the ideal in the "science" of paleontology for the reason that it is not a certain fact that the immediate ancestor to birds went through the process of fossilization. But then Jacobs, in his further comment, assumes too much – a good hypothesis predicts the inevitable find of the unknown ancestor of birds and this eventual find, rest on the powers of comparative anatomy. Jacobs is guilty of subjective reasoning. In a subtle way he is hoping to establish evidence for organic evolution but homology is a long way off from proving common ancestory of birds and coelurosaurian dinosaurs. This is due to the weakness and outdated assertions of comparative anatomy.

I sat at my desk in disbelief after reading what creationists refer to as "paleobabble." The scientific method has definitely changed since I was a young man. Adrian J. Desmond gives us further insight into the formulation of the scientific theory. He writes in *The Hot-Blooded Dinosaurs a Revolution in Paleontology* :

"But, in spite of the inherently conservative opposition to change that characterizes all human activity, science – in contrast to Scripture – is a discipline that is constantly evolving, and is therefore amenable to being updated to accord with the latest findings. Science is not the God-given Truth. It only strives towards a greater understanding of the world, perhaps none greater than when the earth was removed from the centre of it. Scientific theories only approximate more and more closely towards the truth, they will never reach it."

P.145

This is what I understand through Jacobs' words - Theories vary from one scientist to another. One "good" hypothesis is like any other hypothesis so long as it is built on the foundation of logical predictability. On the other hand, Desmond has assumed that theory and hypothesis will lead a scientist toward the truth but his assumption is based on the fact that a Supreme Intelligence is not a part of the scientific picture.

As a creationist, I cannot assume an undue bias towards the Scripture. I have read over a hundred books on dinosaurology and have concluded that researchers have wasted countless man hours and some of them an entire lifetime on proving something which *the* Bible says does not exist – organic evolution. There are many hundreds of creationists who have their PhD's and work in scientific fields. And no, this is not an oxymoron phenomenon! These are scientists who cannot turn their back to the guiding principles divulged by the Creator. With no limiting factors set up by the Creator and with no guideposts to lead a scientist in his search for truth, *how can this same scientist know what truth is or be in a place to recognize a "good" hypothesis from a "bad" one?*

The above quotes demonstrate that the new approach in scientific methodology is sadly lacking. Theories are a dime a dozen and whosoever places confidence in them as leading man down the pathway to truth, is mistaken. Evolutionists claim that theories can be tested in laboratories but *how* is evolution to be tested? The spontaneous development of life from non-living substances or tracing the original ancestors of present life forms back millions of years cannot be tested. The filling in gaps of the geological record merely with the hypothesis that the doctrine of morphology or comparative anatomy is the all-efficient answer to the problem is *not enough* to test the theory of evolution.

I cannot accept the above statements by Jacobs or Desmond as being sufficient guideposts for reasonable man in an understandable world. Rather, I see these ideas as establishing man as a prejudiced and opinionated creature wandering around in an irrational world without guidelines and being tossed about on the seas of theory and blind suppositions.

Jacobs has openly admitted that so far as reflecting transition of species, the fossil record has failed to live up to this ideal. "But, so what?" he responds. He conjectures that the paleotological record is not needed for a scientific opinion and he is not alone in this point of view. Mark Ridley, British zoologist and evolutionist has written:

"….the gradual change of fossil species has *never* been part of the evidence for evolution….In any case, no real evolutionist, whether gradualist or punctuationist, uses the fossil record as evidence in favor of the theory of evolution as opposed to special creation."

["Who Doubts Evolution?" *New Scientist*, vol.90 (June 25, 1981), P.831.

Ridley was in the Department of Zoology at Oxford University]

Anthropologist Vincent Sarich has stated:

"No matter what the creationists may pretend, the fossil record is not, and never has been, our major source of information about evolutionary relationships" (Sarich's quote taken from *Darwin's Leap of Faith* by John Ankerberg and John Weldon, P.227).

The above quotes must be a great source of discomfiture to evolutionists. If Ridley and Sarich were preaching this message from a soapbox in some great Science Hall of Learning perhaps their audience would react in the following manner:

The embryologists would merely snicker; the anatomists would nod with approval; the biologists would snort with laughter; the cladists would jeer and heckle; all gradualist evolutionists would bend over and become hysterical; the punctuationist evolutionists would hiss and boo; the creationists would simply shake their head in disbelief.

It is a wonder that the ideas of Ridley and Sarich could be published with any feasibility and especially in the light of evolutionary dependency upon paleontology that is kept constantly before the public eye. Every high school student is completely and totally aware of the fossil record and its vital importance to the cause of evolution. The magazines (for example the *National Geographic*), newspapers, and journals are frequently printing fossil finds as one of the queenly icons of Darwinism. Even the common person of the streets knows that paleontology is the devil's advocate in defending the progression of life's forms.

There are professionals more candid in their approach to the general problem of transitional life forms in the fossil record. For example, Gregory S. Paul (a writer and artist of dinosaurs) proclaims:

"Especially pertinent to our problem is how the first birds and bats both appear in the fossil record suddenly and fully developed – in neither case are primitive flying grade forms known ... *That the major flying vertebrate groups always appear fully developed so suddenly is very suspicious* ... In the case of birds, we have been lucky to find one of the protoflyers, *Archaeopteryx*. Besides, the distance between *Archaeopteryx* and full birds should not be exaggerated, for it does not take radical and tremendous alterations to make a protobird into a bird." [Emphasis, mine]

Predatory Dinosaurs of the World (A Complete Illustrated Guide), Pp.68-69

Apparently, according to Paul's last sentence, he believes that "evolution can be gradual but it can also move with awesome swiftness." He admits that it is suspicious that major flying groups can appear suddenly and with no transitions in the fossil record. I find his choice of the word "suspicious" suspiciously close to the creationist's prediction of the creation model wherein life's fully developed entities came forth from the hand of God and without any transitions or ancestors leading up to their biogenetic existence.

Here are the candid words of the noted evolutionist Pierre Lecomte du Nouy, an expert in the science of statistical probability:

"We are not even authorized to consider the exceptional case of the *Archaeopteryx* as a true link. By link, we mean a necessary stage of transition between classes such as reptiles and birds, or between smaller groups. An animal displaying characters belonging to two groups cannot be treated as a true link as long as the intermediary stages have not been found and as long as the mechanisms of transition remain unknown."

Human Destiny, P.72 (Quote found in *The Farce of Evolution* by Hank Hanegraaff, Pp.37-38).

Think about the implications of this statement! If reptiles evolved into birds, thousand of intermediate stages should have been discovered but such is not the case. More than this: not one discovery of a form between a reptile and *Archaeopteryx* has ever been found nor between *Archaeopteryx* and a bird. Countless field workers have been combing the fossil cites all over the world and not one of them has ever returned to camp holding within his hand the transitional fossil that leads up to *Archaeopteryx* or that leads away from *Archaeopteryx*. Pierre Lecomte du Nouy does not hold the "ancient-wing" to be proof of evolution as many evolutionists insist. In fact, he does not even consider it to be a link. How can one animal displaying the advanced characteristics of two animals be considered a link? There is no paleontological evidence to authorize such a theory.

The duckbill is like a mammal and like a reptile. Do evolutionists consider this animal to be a connecting link between them? The wildebeest is like an ox and like a horse. No evolutionist considers it to be a connecting link between the two. Because *Archaeopteryx* has teeth and claws like a reptile and wings and feathers like a bird, the supposed half-reptile/half-bird can no longer be used as a proof of evolution.

Henry M. Morris (the creationist) agrees with Pierre Lecomte du Nouy (the evolutionist) and has written:

"However, at the very most, *Archaeopteryx* was a 'mosaic' form, not a transitional form. That is, each of its attributes was fully formed and functional, not incipient or atrophying. Its wings and feathers were complete and perfect, not half-legs or half-scales in the process of evolving into wings and feathers."

Science and the Bible

P.53

Darwin's doctrine of gradualism is contradicted by the philosophical problem of no transitional forms in the fossil record. It is a matter of faith for scientists, who believe in evolution, to advance their notion in spite of the fact that there are no observable mechanisms in nature or in fossil documentation. *Archaeopteryx* cannot serve as the watertight argument that Darwin hoped it would be. How can a specimen serve as the transition between birds and reptiles when it offers no precursor of the crucial scale-to-feather or leg-to-wing transformation?

How can paleontologists accept an animal as an intermediate when it does not display the proper anatomical structure between two animal classes? Gish writes:

"With specific reference to …. *Archaeopteryx*, creation scientists point out that research during the past few years on major anatomical features of *Archaeopteryx* had established, in every instance, that these features are bird-like rather then reptile-like, and thus its status as a transitional form is becoming more and more dubious with the passage of time."

Creation Scientists Answer Their Critics

P.298

Archaeopteryx is not the incontrovertible proof for the origin of birds as emanating from coelurosaurian dinosaurs or any other theory being circulated at this present time. The missing links that are observed on a global scale are more than enough to establish affirmative support for Bible creation. This is why the fact of discontinuity among fossils with the complete lack of intermediates is crucial to the question of creation versus evolution. Among other things, the theory of evolution asserts that the beings now living have descended from different beings which have lived in the past. Without transitional forms, this definition is neither feasible nor within the boundaries of rational reasoning. Also, evolutionists have to explain how natural forces make it possible for living forms to arise spontaneously. As a creationist, I would direct the reader's attention to the words of Frank Lewis Marsh who succinctly placed the entire issue before us:

"The only possible way to see evidence of evolution in the fossils is to be first completely sold to the idea of evolution. This point of view appears to furnish evolutionists with the vast amount of credulity necessary to bridge the innumerable gaps between kinds, and to make them oblivious of the true significance of the entire lack of intergrading forms. On the other hand, the first eight chapters of Genesis present an account of origins which *adequately* explains *every* fact of the fossil record." [Marsh emphasis]

Evolution, Creation, and Science

P.368

At the beginning of this chapter, statements from certain scientists demonstrated their disdain for the fossil record and their belief that it was not consequential in establishing the theory of evolution. Other evolutionists were much more candid in setting forth the shortcomings of paleontology and its failure to provide the essential transitions and intermediates that were admittedly indispensable to the doctrine of ancestral origins. Still other scientists must be heard from simply because their admissions will help us to understand that the paleotological record is, after all, not only essential to the theory of evolution but without it, evolution would appear as an insignificant and trivial tenet.

Steven M. Stanley writes:

"While many inferences about evolution are derived from living organisms, we of History of Science University must look to the fossil record for the ultimate documentation of large-scale change. In the absence of a fossil record, the credibility of evolutionists would be severely weakened. We might wonder whether the doctrine of evolution would qualify as anything more than an outrageous hypothesis." "*Macroevolution: Pattern and Process*," P.2 (San Francisco: W.H. Freeman and Co., 1979, 332 pp).

David B. Kitts writes this revealing comment:

"An examination of the work of those paleontologists who have been particularly concerned with the relationship between paleontology and evolutionary theory, for example that of G.G. Simpson and S. J. Gould, reveals a mindfulness of the fact that the record of evolution, like any other historical record, must be construed within a complex of particular and general preconceptions not the least of which is the hypothesis that evolution has occurred" ("Search for the Holy Transformation," review of *Evolution of Living Organisms*, by Pierre-P. Grasse, *Paleobiology*, vol.5, summer 1979, pp.353-355).

Dr. Steven M. Stanley has received various awards for his work in paleontology. He openly admits:

"The known fossil record fails to document a single example of phyletic [gradual] evolution accomplishing a major morphologic transition and offers no evidence that the gradualistic model can be valid" (Stanley's quote is from *Darwin's Leap of Faith* by John Ankerberg & John Weldon, P.217).

The facts of nature revealed in the science of paleontology simply do not fit the model of evolution. What other recourse does one have except to see if the facts fit the model of creation? [When a phenomenon is difficult to observe directly, scientists use predictions to carry out a working hypothesis. This is one way of reaching a logical and feasible conclusion. For example, an evolution model of paleontology would predict a gradual change of simple forms into more and more complex forms. It would also predict that transitional series would link up the various classes of species with the absence of systematic gaps. On the other hand, a creation model would predict a sudden appearance of fully developed and specialized forms. These forms would be highly complex and with sharp boundaries separating major groups. It would also predict that there would be no transitional forms existing between distinctly created types. On the one hand, creationists base their model on empirical evidence. On the other hand, evolutionists make use of a model that is totally void of confirmation and, therefore, their evolutionary hypothesis fails the test of empirical science]

In the light of what we have learned about the fossil record and its inconsistencies in presenting a substantial case for evolution, it is quite a shock to read the following statement:

"Only fossils can tell us the pathways that evolution did take, in addition to where they ended up. And it is only through the fossil record that we can observe, empirically, the results of the admixture of earth and life processes."

Quest for the African Dinosaurs

Louis Jacobs

P.26

Jacobs professes to be an expert in paleontology and dinosaurology. In our study, thus far, we have observed that *fossils do not mark the pathway of evolution* and cannot be considered, in the slightest aspect, to give empirical information that backs evolution. Such a statement has been extricated from the dream word of paleontology and not from its reality.

From a comparison of the two models, in spite of Jacob's comment, one must agree that the creationist model best fits the facts of nature and paleontology. The fossil record is not spotty and inadequate. There have been countless millions of samples collected from all over the earth and gaps are not the result of faulty sampling. The gaps are real and the lack of transitions between classes is factual. Even Punctuated Equilibrium is not a good explanation of the lack of intermediary fossils. As Richard M.Ritland has written:

"Is it plausible that 'blind fate' would *always* miss recordings such transitions between higher categories and yet preserve abundant remains of the basic stable basic types? Perhaps the most natural explanation is that the missing links between major groups never really exited." [Emphasis his]

A Search for Meaning in Nature

P.153

Of all the points made concerning transitions, this short thought is perhaps the strongest I have ever read by a creationist. From the above statements by Stanley and Kitts, evolutionists appear to be in a quandary so far as being persistent in their attempt to build a substantial defense of their theory. Some evolutionists deny that "missing links" are a major problem to the theory of organic evolution; simply come up with a "good" hypothesis that predicts what the future fossil finds will be and this will be good enough to keep evolutionary doctrine intact. But this is reading into the fossil record a phenomenon which has never been observed in the past, the present, and only imagined for the future. The fact remains there are no transitional forms between phyla, none between classes, orders, or families of extinct life structures. Such intermediates can not be found anywhere on the face of the globe. *Archaeopteryx* and other alleged transitions are "mosaics" and not true transitions. Why should some evolutionists have any hope that their "good" hypotheses of prediction for fossils will one day come to pass? The hundred thousand fossil species discovered since Darwin's time have not fulfilled the evolutionists' dream of "missing links." Why should the future be any different from the past?

Other evolutionists still place their confidence in paleontology as the only science that designates macroevolution. In other words they look to the fossil record for the ultimate documentation of large-scale change. However, the imperfection of the geological column in providing transitional forms of life has thwarted their purpose for mapping out the path of organic evolution.

Still other evolutionists have attempted to explain the lack of "missing links" through Punctuated Equilibrium. We have already noted some of the serious objections to this theory in chapter three.

The general problem of transitional forms in the paleotological record is *fossils are not transitional*. Many species, at the various stages in the geological column, are already in an advanced state of "evolution" the first time they appear. Many evolutionists admit that such species come into view without any evolutionary history. That is, they appear suddenly and in full perfection. Thus, the major problem for evolutionists and is a crucial factor in the undermining of their theory, can be summarized in the following sentence: When it comes to a study of the fossil record, life forms seem to have taken their place not by *transmutation* but rather by *substitution*. **The indisputable and incontrovertible fact when it comes to our present study of *Archaeopteryx*, there is no stratigraphic proof for a Darwinian origin of birds. The facts are that all life forms whether they are living or extinct fall into existing major groups and without any transitional structures leading up to them.**

Do the paleontologists have other recourses of study? Are there alternative ways for them to advance the notion that the *Archaeopteryx* can be traced in its timed sequence throughout the geological column from the Permian strata all the way up to the Cretaceous-Paleocene levels? This chapter has addressed the issue of the general problem of transitional forms. Most scientists, who deny that transmutation of the species can be revealed only through fossil forms, have turned to other sources for more specific ways to express their belief in organic evolution. The following chapter will concentrate on these more specific issues but I shall make an attempt to lead the readers into seeing how these specifics can be more of a hindrance to the paleontologist rather than an aid in his efforts to advance the cause of evolution.

CHAPTER 5

THE TALE OF TWO QUESTS -

BIOCHEMISTRY AND CLADISTIC ANALYSIS

(Do these Two Methods Solve the Problem of Fossil Transitions?)

"But ask the animals, and they will teach you, or the birds of the air, and they will tell you; or speak to the earth, and it will teach you, or let the fish of the sea inform you. Which of all these does not know that the hand of the LORD has done this? In his hand is the life of every creature and the breath of all mankind."

Job 12: 7-10

NIV

The general problem of the paleontological record – the lack of intermediates – has been a long going predicament to evolutionary scientists. The following statement by George Gaylord Simpson not only describes this problem comprehensively but also describes what it has done to disillusioned naturalists:

"On still higher levels, those of what is here called 'mega-evolution,' the inferences might still apply, but caution is enjoined, because essentially continuous transitional sequences are not merely rare, but they are virtually absent. These large discontinuities are less numerous, so that paleontological examples of their origin should also be less numerous; but their absence is so nearly universal that it cannot, offhand, be imputed entirely to chance and does require some attempt at special explanation, as has been felt by most paleontologists ....

"As it became more and more evident that the great gaps remained, despite wonderful progress in finding the members of lesser transitional groups and progressive lines, it was no longer satisfactory to impute this absence of objective data entirely to chance. The failure of paleontology to produce such evidence was so keenly felt that a few disillusioned naturalists even decided that the theory of organic evolution, or of general organic continuity of descent, was wrong, after all."

Tempo and Mode in Evolution, Pp.105-106,115 (Simpson's quote is in *Life, Man, and Time* by Frank Lewis Marsh, P.173).

Disillusioned naturalists may have forsaken the theory of organic evolution but other scientists have turned to what they believe to be more formable explanations for their theory. They seek out confirmation in the various branches of biology – avenues that they may turn to in time of trouble. But these ways, as we shall soon discover, will often lead to more problems than the evolutionists' attempt to solve the old one of gaps in the fossil record.

Let us tackle the first specific item that scientists employ to overcome the problem of no transitions in the fossil record – biochemistry. Because the fossil record does not offer to evolutionary scientists the picture of graded series of fossil organisms over long periods of time, the evolutionists have high hopes that what the fossil record fails to do, biochemistry will provide the evidence of gradual change between taxonomic groups. One of the purposes of this chapter will be to determine whether this hope has been realized. The approach will be to first establish the faith that evolutionary scientists have placed in this method; find out the ways they have of deciding on how to preserve the chemical content of certain fossils; discovery what the chemistry reveals about these same fossils; to finally settle the issue if biochemistry is a potent force to establish the validity of evolution or serves to weaken the foundation and superstructure of Darwin's theory of progressive advancement of species and his branching-tree pattern of evolutionary relationships.

Since this book's main focus is on the *Archaeopteryx* and since so many evolutionary scientists and especially in the field of paleontology believe that birds evolved from dinosaurs, the main examples of biochemistry will involve dinosaurs and dinosaurologists. John H. Ostrom wrote:

"Were it not for the feather impressions, *Archaeopteryx* would never have been classified as avian, but have been labeled dinosaurian." "Reply to 'Dinosaurs as Reptiles,'" *Evolution*, vol.28 (September 1974), P.493

Ostrom believed that *Arxchaeopteryx* evolved from small theropods (small dinosaurs with serrated, saberlike teeth). Ostrom's views are highly debatable and his theories will be challenged in chapters to come but at this present time and for the sake of argument we will make use of dinosaur examples for biochemistry and its significance to evolutionists. The chemical analyzation of fossil bones is important to the evolutionist for substantiating his theories. Whether the bones are of dinosaur species or other species it matters not to the biochemist since his analyzation of the specimens will have the same alleged results – establishing the assumed linkage between classes of animals.

The subject of preservation of chemistry in fossil bones was at one time ignored but more and more paleontologists are becoming interested in the taphonomic history of fossils. These taphonomists study fossil forensics – what happens to an organism from death until long after burial. At first there was an emphasis on postmortem and preburial history but now there is a widely-circulated interest on postburial history of fossils.

Incredible detail of fossil bones placed under a microscope reveals individual bone cells and often with the original cells still intact – unchanged and not replaced (replacement is the process of fossilization involving substitution of inorganic matter for the original organic constituents of a plant or animal. This definition is according to the *Dictionary of Geological Terms*, third edition prepared by The American Geological Institute, 1976). This is vitally important for determining that at least small traces of the original composition of the fossil have remained in tact and is, in fact, the original composition. This detail gives evolutionists the hope than eventually proteins will be analyzed through amino acid sequencing. In 1989 a Bone Chemistry Workshop was set up in Los Alamos National Laboratory (Los Alamos, New Mexico). The first recognition of organic molecules from a dinosaur was established with all its evolutionary implications. David D. Gillette, a Utah state paleontologist, writes:

"With our discovery of noncollagen protein in a dinosaur bone comes many implications for future research on fossil bones. The potential applications are almost overwhelming. For example, we could begin to use comparative biochemistry of proteins and other biological macromolecules to test phylogenitic reconstructions of the various kinds of vertebrates – and thus determine who is related most closely to whom. We could better our understanding of the evolution of biological macromolecules, such as collagen and various bone proteins ....And if there is protein, perhaps there is nucleic acid too. The futuristic science underlying the book and film *Jurassic Park* may not be all that far-fetched."

P.149

Finding a dinosaur bone that had once been buried in the earth and then, through erosion, exposed on its surface and long enough to be discovered by a paleontologist, is one chance in a million. There are even drilling devices used to preserve bone tissue from contamination so that the samples can be analyzed with more accuracy. But Gillette has gone overboard in his enthusiasm. Like other paleontologists, he has jumped on the bandwagon of new possibilities for animal comparisons in the fossil record in order to substantiate evolutionary theories but as "Jurassic Park" is filled with more science fiction than science truth, Gillette's vision of "comparative biochemistry" for phylogenitic reconstructions may prove to be "far-fetched" after all. One other scientist should be mentioned. W. Dale Spall, a geochemist at Los Alamos National Laboratory claimed in 1991 that he had isolated proteins from a backbone of Seismosaurus.

From David D.Gillette, Spall, and The Bone Chemistry Workshop in New Mexico we will go to the area of Alaska and note some other scientists who support biochemistry as the answer to the perplexing problems posed in the doctrine of progressive paleontology. Philip Currie claims that the reason the dinosaur fossils of Alaska are so remarkable is because instead of being their normal weight of a rock's heft, they are "as light as dog bones a few days old." Philip Currie is a Canadian paleontologist who, among other things, has hunted in the dinosaur bone beds of Alberta in the late 1980s.

Gordon Curry (don't confuse him with Currie!) and his collaborators have claimed that the Alaskan rocks contain fossils that are far from devoid of life. The extinct life forms contain remains of living tissue. They maintain that:

"These biomarkers have broken down into their constituent amino acids, but are still sufficiently preserved to enable scientists to track molecules to their origins ….The sequence of those amino acid molecules differs from individual to individual as well as from species to species. DNA holds the genetic code, but is locked in soft-tissue cells, seldom preserved in recognizable fossils… DNA may be hard to come by in fossils, but amino acids can now be detected in fossils from almost incomprehensibly small amounts."

Don Lessem

Pp.222-223

Gordon Curry believes that should he find Alaskan dinosaur proteins, he would be enabled to sort out evolutionary pathways according to similarities in molecular structures between different dinosaur groups. He might also tie down the rate of evolution and how quickly dinosaurs were changing per million years.

All these claims sound good on paper but they fall short when the real facts are investigated. As usual, the evolutionist makes bold assertions when anything new is thought to uphold the tenant of evolution. He attempts to get the public to accept his over-emotional statements and then sits back in satisfaction with another job well done. Says Curry, "This is totally speculative, but with the sequence of proteins in their DNA you could regenerate a piece of dinosaur, a very small part of a molecule, but a dinosaur's nonetheless."

[Ibid. P.223]

Few of the public know just how tentative Curry's claim is! Not only is the immediate statement above totally speculative but everything he has said up this point is unproven and completely unfounded. As far as his being able to trace the ancestral pathways of dinosaurs into their various groups and even tracing the rate of evolutionary change per million years is as the British would say, "Mere rubbish and poppy-cock." Just as "Jurassic Park" was bad science but good entertainment so the words of evolutionists and their use of biochemistry also makes for bad science but features good science fiction.

Before speaking to these issues, let us observe what many scientists are actually saying. First of all, through biochemistry, evolutionists are able to determine the linear sequence of genes now thought to originate in DNA and are responsible for the transmission of hereditary characteristics. This is all well and good thus far. However, evolutionists believe that genetic change not only programs the individual's development but generations of accumulating genetic changes cause new species to evolve.

Also, in the second place, any complex organisms can be traced back through their common ancestors to one or several simple organisms. The theory that animals and plant types have their origin in other preexisting types and the distinguishable differences are due to modifications in successive generations is called organic evolution.

Let us tie in these thoughts to the subject of this book – *Archaeopteryx*. We know that DNA is the fundamental chemical in life. It is a complex molecule found in all life forms and is the "blueprint" that relates to an organism how to build itself. Most evolutionists have turned from the fossil record because of its gaps and lack of intermediary transitions and also because homology (similarity of body parts) failed to prove evolution. [For those readers who are interested in seeing how homology and biochemistry failed to prove evolution see my first book – "Was Darwin Wrong? – YES" and especially chapter four: "The Giraffe's Neck – The Big 7" (homology) and chapter eleven "The Orangutan's Visit to Dodger Stadium" (biochemistry).

Evolutionists have turned to biochemistry since they allege that DNA maps out the history of relationships among organisms. In this way fossils are once again, through biochemistry, providing evidence in support of evolution by tracking the progress of life forms as they change through time. In *"Rocks, Fossils, and Dinosaurs"* Christopher A. Brochu and Colin McHenry, in chapter four, give the classic example of the *Archaeopteryx* being the link between the two groups of birds and dinosaurs. These writers also suggest DNA sequences have something to do with *Archaeopteryx* – at least in their inference. Writing about *Archaeopteryx* as a small dinosaur and how the molecule DNA controls the form of all organisms, the statement is made:

"The similarity of body parts (homology), the similarity of DNA sequences, and the changes in the fossil record make sense only in an evolutionary perspective."

P.90

The writers plan this statement to mean that such facts can be explained only in an evolutionary perspective or the facts would otherwise be inexplicable. But I am going to be facetious and take the statement against their intention by saying that such facts can be explained *only by evolutionists*. Otherwise, they make little sense to anyone else familiar with the true facts and that has a proper perspective of nature. My remarks are not for the purpose of being funny or frivolous. I consider the above statement by Brochu and McHenry to be at the height of dogmatism.

What they are stating in so many words – "Nothing in biology makes sense except in the light of evolution." Henry M. Morris writes (responding to such dogmatic statements by evolutionists and how evolution has achieved its position):

"Amazingly, however, it [Evolution] has achieved this position of intellectual eminence without on iota of real scientific proof. It has become the world view, the ruling paradigm, of every discipline of study. …. It maintains this control, not by objective scientific argument but by the anti-creationism of its leaders and spokesmen."

That Their Words May Be Used Against Them

P.107

We are now ready to consider if biochemistry really maps out the tracking of life forms as they change through time. Does DNA map out the history of relationships among organisms that have been fossilized long ago? Is Darwin's theory that the origin of new species comes from previous existing species substantiated by biochemistry? Just how important is the trail of "species" to the evolutionists? Jonathan Wells writes concerning the "smoking gun" of evolution, which is speciation:

"Darwinism depends on the splitting of one species into two, which then diverge and split and diverge and split, over and over again. Only this could produce the branching-tree pattern required by Darwinian evolution, in which all species are modified descendants of a common ancestor.

"So if biologists could start with one species and make it split into two by using variation and selection, they would be able to observe branching speciation, the cornerstone of Darwinism. They would have found evolution's smoking gin."

The Politically Incorrect Guide to Darwinism and Intelligent Design

P.51

Again, we ask the question – how important is the trail of "species" to the evolutionist as he maps out the branching tree or bushes in his endeavor to determine the evolutionary history of organisms? Obviously it is vital. Otherwise, to the evolutionist, Darwin's origin of species would remain unsolved and unproved. After all, the theory of evolution is supposed to solve these origin dilemmas; not create new ones.

Because evolution is supposedly an ongoing process, most creationists realize that evolutionists have different definitions of species. According to some evolutionists, there are more than twenty-five definitions. However, let's talk about those species concepts which are used in taxonomic units for identification and classification for extinct animals.

When for example, dinosaurologists write that *Archaeopteryx* can be traced back to theropod dinosaurs or that this fossil was transitional between dinosaurs and modern birds; such claims can only appear to be preposterous to the creationist. There are many Evolutionary Trees, depicted in books written for the public, tracing the *Archaeopteryx* all the way back in deep time through the Dinosaurs, Thecodonts, Archosaurs, Diapsids, and through the Ichthyostega of the geologic record (this tracing through the geological column will be elaborated upon in chapter twelve). **When these same dinosaurologists write that DNA can map out the history of organisms that for millions of years have been extinct, they must be writing from the standpoints of hope for the future or using biochemistry and its application on extant animals to somehow become relevant for extinct animals.**

In other words, they must deem the present findings of biochemistry on living animals and the alleged results of their studies to be applicable to extinct fossilized animals of the past. When the evolutionary scientists apply genetics and behavioral traits to fossils, we wonder what species definitions best fit these relics of nature. Genetics, natural selection, and Linnaeus's definition of a species simply do not fossilize as translatable information. Linnaeus defined a species as a group of organisms capable of freely interbreeding. This is a position rarely contested by present day scientists.

Michael Novacek, Curator of Vertebrate Paleontology, at the American Museum of Natural History in New York City, writes about the millions of biological entities (species) put into some kind of organizational chart as the result of biological classification. He mentioned John Ray of England and Carolus Linnaeus of Sweden (both creationists) as "arranging life forms into lists based on very explicit characteristics from every part of the anatomy." [These lists and charts were taken over by the evolutionists who adapted the Linnaeus nomenclature for species and added tons of anatomical comparisons to demonstrate the relationships of organisms. But they also evaluated them in the "light" of evolution to indicate that one or several simple organisms changed into more complex organisms; also called organic evolution] Novacek contends that there are some biologists who claim that the best criterion for recognizing a species is to go by the Linnaeus definition. Novacek states why biologists have a hard time dealing with species and especially when Linnaeus's definition is applied to life forms of the fossil record:

"Naming species, then, may seem straightforward; *Tyrannosaurus rex* is not likely to be confused with *Rosa gallica* [the French rose]. Oddly enough, biologists have a hard time dealing with species as a general phenomenon. There is a continuing debates over which are the best criteria for recognized species. Some biologists favor the notion that species are best recognized as populations that are reproductively isolated (do not interbreed with members of other populations to produce fertile offspring). Unfortunately, most objects in the biological world are not like our family roses; we cannot examine directly their capacity for interbreeding, when it comes to a fossil that is certainly true. We cannot mate a *T. rex* with its Mongolian relative *Tarbosaurus bataar* in order to see whether it is the same or a different species. There are other complications...."

Dinosaurs of the Flaming Cliffs

P.64

It is understandable how evolutionists seek to establish, through homology, the evolutionary history of organisms in the fossil record. The similarity of body parts at least allows a basis for the evolutionary philosophy of a common ancestor. This old way of determining how dinosaurs are related to one another (homology) by studying their structural comparison seems, at the very least, coherent. But the new method of determining phylogenitic relationships of animals by their molecules, especially dinosaurs, is mind-boggling. For extant animals, molecular phylogeny must assume Darwinism to be valid and then fit the data into a branching-tree pattern. For extinct animals, not only must Darwinism be assumed but it entails the further hypothesis that the molecular data of present living animals can be projected into the past and applies to dead or fossilized animals. **In the first place, molecular studies have failed to produce a consistent evolutionary tree for living forms. In the second place, how are molecular studies**

carried out on fossils with no DNA patterns supposed to establish common ancestors?

I attended a creationist meeting a few months ago and was told by the guest speaker that an unfossilized *Tyrannosaurus rex* bone was discovered with evidence of hemoglobin and red blood cells. I was led to the conclusion that this discovery invalidated the deep time of millions of years and radiometric dating methods. Although believing in a "young earth" I had my doubts that dinosaurs died so recently.

Mark Isaak writes that Schweitzer and her colleagues found the bone and it was "not completely fossilized." He adds, "But lack of permineralization does not mean unfossilized." He also writes: "Schweitzer et al.did not finds hemoglobin or red blood cells. Rather, they found evidence of degraded hemoglobin fragments and structures that might represent altered blood remnants. They emphasized repeatedly that even those results were tentative, that the chemicals and structures may be from geological processes and contamination..."

The Counter-Creationism Handbook

P.141

It is hard to agree with his assessment of the hemoglobin fragments. In reading the above Handbook, with most of Isaak's comments I do not agree. His commentary on "Biblical Creationism" is ludicrous; sections on Biology and Paleontology have been contradicted by other evolutionary scientists; but in other sections he has brought in arguments that non-professionals would find difficult to answer.

In the above reference, Isaak comes across as fairly reasonable in his evaluation of the *T. rex* bone. He read Schweitzer's report and merely stated what she and J. R. Horner had to say about the bone in 1999. [See the report of M.H. Schweitzer and J.R. Horner. 1999. Intravascular microstructures in trabecular bone tissues of *Tyrannosaurus rex*. *Annales de Paleontologie* 85; 179-192]

However, Schweitzer did say to a lab technician: "The bones, after all, are 65 million years old. How could blood cells survive that long?" [ACTS & FACTS, October 2009, Dinosaur DNA Research, James J. S. Johnson, Jeffrey Tomkins, and Brian Thomas]

If she said this, then empirical science – by her own observation – dictates that she is completely wrong in her old-earth evolutionary theory. This would set UNIFORMITY back on its heels and make the YOUNG-EARTH THEORY – in fact, an accurate, empirical truth.

What about W. Dale Spall – the biochemist who claimed that he had isolated proteins from a backbone of *Seismosaurus* in 1991? And what about Gordon Curry and his collaborators who maintained that Alaskan rocks contain tissue of dinosaurs and amino acids can be detected?

The significance of W. Dale Spall isolating proteins from a dinosaur, according to evolutionists, has a high impact on their theory. According to Gordon Curry, the finding of dinosaur proteins would give him the capabilities of tracing the pathways of dinosaur groups. Curry hopes that proteins will be analyzed through amino acid sequencing. Through such biochemistry he hopes to test phylogenitic reconstructions of the dinosaur vertebrates and determine who is related to whom.

In comparing statements written by other molecular biologists, Curry's claim of tracking dinosaur pathways, borders on the unattainable. Michael Denton has written:

"The really significant finding that comes to light from comparing the proteins' amino acid sequences is that it is impossible to arrange them in any sort of evolutionary series. ....

"Thousands of different sequences, protein and nucleic acid, have now been compared in hundreds of different species but never has any sequence been found to be in any sense the lineal descendant or ancestor of any other sequence."

Evolution: A Theory in Crisis

Pp.289-290

Spall's and Curry's claims in tracing ancestral pathways are highly speculative but let's assume that these scientists had all the biochemistry information they needed – that is, they had all the necessary amounts of DNA samples for tracing dinosaurs. Would scientists be lead to believe what could not be accomplished in the land of living organisms can suddenly be consummated in the land of the dead? If molecular evolution fails to confirm the hypothesis of neo-Darwinism and substantiate the branching tree pattern for species of the extant variety (organisms which live in the present), then biochemistry can not possibly validate organisms of the past (for example, dinosaur fossils) as falling into any pattern or pathway of ancestral relationships.

Christian Schwabe, "On the Validity of Molecular Evolution," *Trends in Biochemical Sciences* (July 1986) has written:

"Molecular evolution is about to be accepted as a method superior to paleontology for the discovery of evolutionary relationships. As a molecular evolutionist I should be elated. Instead it seems disconcerting that many exceptions exist to the orderly progression of species as determined by molecular homologies; so many in facts[s] that I think the exception, the quirks, may carry the more important message."

The facts are the mysteries of the DNA-RNA complex are becoming so involved to many molecular and genetic biologists, that the inheritance of homologous structures from a common ancestor has become a rather hopeless assumption. It is becoming clear to scientists: the more analyzation of molecules from various species of animals, the more elusive Darwin's tree or bush of life becomes.

The other night, I made my usual weekly trip to *Barnes and Noble's* book store. The science section had all kinds of new books on Evolution. This was not unusual but what was odd, the word "Evolution" stood out in every title. Nowadays, there are so many books on the market refuting the doctrine of evolution. I would suppose the doctors of evolution have decided that the public needed other dose of tranquilizers to keep them medicated against the Headaches and Unrest caused by creationists and ready-to-step-across- the-line evolutionists. In a perusal of these books, I couldn't help but notice the same old arguments accompanied the icons of evolution. Especially when considering the new scientific pet of evolutionists – molecular biology.

The claim is still made that molecules tell us all we want to know about our evolutionary ancestors and that we can rest assured that we have a common ancestor or ancestors. **But no matter how many books are published, the hypothesis about past evolutionary relationships of organisms being known from molecular data found only in living species, because of its inaccuracies and uncertainties, can no longer be considered a source of pride to its practitioners.**

As the evolutionary biologists Martin Jones and Mark Blaxter have stated: "The true relationships of the major groups (phyla) of animals remain contentious."

Jonathan Wells writes:

"So the molecular phylogeny of the major animal phyla is a mess. When biologists use molecules to look for the trunk and root of Darwin's tree of life, the mess gets worse."

The Politically Incorrect Guide to Darwinism and Intelligent Design

P.43

The faith that many evolutionists have placed in this method of molecular phylogeny has been misplaced. One of the bizarre deceptions of "science" is the reading of a universal common ancestor into the molecules from different organisms. The contradictions resulting from this method of analyzation should be enough to turn every evolutionary scientist away from forcing their studies to fit the mold of Darwin's tree of life.

Jonathan Wells writes:

"…one thing is clear; molecular phylogeny has failed, utterly and completely, to establish that universal common ancestory is true. The molecular evidence, like the fossil and embryo evidence, is plagued with inconsistencies, and Darwinism must be *assumed* in order to explain it; or, as is often the case, explain it *away*." [Emphasis his]

[Ibid. P.46]

[On the failure of the fossil record to establish common ancestries, please refer to the previous chapter of this book. For some insight on the inconsistencies of embryo evidence, see my first book "Was Darwin Wrong? Yes" Appendix II - "A Short History of Embryology – The True and False Stages."]

There is another hope that many evolutionists (but not all) rally behind; the new method used to determine an organism's family tree which is called cladistic analysis. Since biochemistry has

already been dealt with, this is the second specific item to be considered. The *older* system of classification was developed by Swedish botanist Carl von Linne (1707-78); his system categorized plants and animals by their overall similar characteristics. But the *new* method of classification gives each characteristic an assigned code and then is added to a computer database. In view of the fact that a computer is used, this method is either fool proof in finding evolutionary ancestors or it sends researchers on *a fool's errand* (Keith Thompson). We will investigate why this new method is a cause for controversy in the camp of evolutionists.

There is one point to consider right from the start– the cladistic method is not a new concept except in its use of computers and their producing what looks like family trees. In theory, one of the alleged facts they discovered, after sorting out the information was the sharing of 132 characteristics of birds with dinosaurs. This was supposed to be "hard evidence" that birds actually have descended from dinosaurs but many scientists were left unconvinced. Apparently, the "hard evidence" can be debated.

Carl von Linne (Linnaeus) was a creationist [he was known as "God's Registrar"] who in his book *Systema Naturae* of 1735 attempted the Herculean effort of classifying all of the known organisms on the earth. Taxonomy (also called systematic classification) is a method of describing, naming, and classifying organisms. Linnaeus realized that by using similarity of structure, the species could be categorized together in large groupings called taxa. The groupings grew larger and with the more general traits, the higher up the hierarchy they occurred. As Michael Denton stated, "the underlying hierarchic order of nature was increasingly reaffirmed by nearly all the great naturalists and biologists of the time (with one or two exceptions such as Lamark)."

The hierarchies and nested sets of groups were first discovered by Linnaeus in the eighteenth-century. Many evolutionists accredit Darwin as the first to discover the hierarchical pattern of nature but this is an unbelievable postulate since Linnaeus preceded Darwin by a hundred years. True it is that Linnaeus used a system of folk classification that extends back to the time of Aristotle but nevertheless, he should receive the credit since his system was more efficient and extensive in identifying the basic patterns in nature. Also evolutionists have the nerve to claim that the natural system of classification was discovered by Darwinian taxonomists in the 1950's, based on the predictions of evolution. However, it was Linnaeus who taught in 1735 the nested hierarchy mirrored the divine plan of creation. In other words, he saw these groups as predicting creation.

In response to this nested hierarchy, Darwin surmised in the middle of the nineteenth century, it was the result of the branching-tree pattern with all its life's forms originating from a common ancestor.

In the 1950s Willi Hennig, a German insect worker, considered the problem of how animals are related to one another. He realized that some animals may not look very similar but must be related by common descent or genealogically.

Jonathan Wells gives us insight into Henning's method:

"In Henning's approach, organisms are simply assumed to be related by common descent, and their characteristics are then used to infer the points where their lineages diverged into separate branches (hence the name, 'cladistics').... ....

"The order in which animals appear in the fossil record also becomes secondary or irrelevant. If evolutionary relationships are inferred solely on the basis of character comparisons, an animal can be the descendant of another even if the supposed ancestor doesn't appear until millions of years later. The fossil record is simply re-arranged to fit the results of cladistic analysis."

Icons of Evolution Science or Myth?

P.119

Cladistic analysis is not easy. Scientists have to study an overwhelming amount of anatomy. When the cladist for example places a dinosaur in a family tree, he has to know every facet of the particular dinosaur's anatomy plus the other species around it.

This has been called "the curse of cladistics" by Jacques Gauthier who is of the California Academy of Sciences in San Francisco. He claims that in order to determine the direction of evolutionary change in groups you are investigating, you will be lead "to quite a structured tree and ever-widening areas of investigation [the curse]...." (*Hunting Dinosaurs* by Louie Psihoyos with John Knoebber, P.123).

Critics of the new method of classification argue that structural comparisons based on similarities called characters or features are not necessarily indicative of common ancestry. Similar designs of species mean they could have risen from the same stock. But there are other ways known to evolutionists. The cladistic analyses could be carried out on species which may have evolved independently. That is, two animal species under similar selective pressures could develop similar adaptations autonomously. This is called convergent evolution; the acquisition of similar characteristics by animals of different groups and distantly related. If two closely related taxa do it, taxonomists call this parallelism. There are other problems in setting up a cladogram or "family tree" and this recently developed methodology inspires controversy on how it should be practiced. For one thing, cladistic principles have no simple formula and it has already been mentioned how complex the method is. Thus every cladist uses what he deems to be the best approach. Some look at two few characteristics of species and come up with weird, wild, and wispy results. Some look at so much anatomy that it is difficult for the human mind to handle it all or to take into account the significance of the findings.

Gregory S. Paul makes this statement about cladistics:

"The problem with cladistics, and all anatomically based phylogenics, is that they are often contradicted by DNA-DNA hybridization studies. This suggests that even the best cladograms may be grossly in error. Then again, it is possible that the DNA-DNA studies are wrong or misleading. If the subjects are long extinct, like dinosaurs, there is no way to compare the two methods. What we can say is that cladograms map character patterns, not necessarily true relationships. But there are many cases where we can show the potential relationship. Even an ancestor-descendant one, has a high probability of being real. This is most true when a series of very similar species is found in a sequence of sediments lying directly one atop another, and the species show a consistent trend of change. However, such straight 'lineages' are the exception. Mostly we are dealing with complex branching patterns or mainstreams of ancestory and descent ... .

"Along with figuring out the relationships of animals, one has to label them and their groups. Sad to say, dinosaur taxonomy has been near chaos, with every one doing pretty much their [his] own thing."

Predatory Dinosaurs of the World A Complete Illustrated Guide

P.173

Cladistics has been called the "cornerstone" of evolutionary analysis but creationists see this claim as contradictory to the course of nature. Since the use of cladistics does nothing more than Linnean systematics which is a type of classification committed to the innate hierarchical pattern of highly ordered groups within groups and to the absence of sequence in nature, cladistics can hardly be a tool for evolution. **Nature shows the reality of discontinuity and the unreality of ancestral pathways. There is no place for evolutionary biology in cladistics. In fact, the real purpose of the cladist is to determine the order of nature in such a way as to free him from *a priori* evolutionary bias. According to evolutionists, this is the standard, which taxonomists have set for themselves.**

However, the doctrine of evolution has set a precedence that prevails over all good intentions. In spite of evolutionary vows to be free from bias, cladistics enables the cladist to visualize those imaginary links in the fossil record. **The real empirical evidence of discontinuity is swept under the carpet of sequential arrangement. The house of evolution is purported to be swept clean of bias and the unsullied notion of true science is presented to the public. But how does true science not recognize the universal feature of all organisms as a part of the hierarchic order of nature, fail to notice the complete absence of intermediates that would bespeak of sequential relationships, read into the fossil record hypothetical ancestral pathways, cry louder than its own evolutionary zoologists who find it unfeasible to relate major groups of life forms in sequential alignment, fail to perceive that random processes could not possibly generate such a highly intrinsic pattern of nature, and finally to overthrow the final analysis nature's order is not sequential.**

The journal *Nature* considers cladism to be a threat to evolutionary biology. This prestigious journal launched a verbal attack against the staff of Natural History in South Kensington when they reorganized the public galleries to accommodate cladistic standards.

Duane T. Gish quotes Patterson, one of the editors of *Nature*:

"But as the theory of cladistics has developed, it has been realized that more and more of the evolutionary framework is inessential, and may be dropped. The chief symptom of this change is the significance attached to nodes in cladograms. In Hennig's book, as in all early work of cladistics, the nodes are taken to represent ancestral species. This assumption has been found to be unnecessary, even misleading, and may be dropped. Platnick refers to the new theory as 'transformed cladistics' and the transformation is away from dependence on evolutionary theory."

Creation Scientists Answer Their Critics

P.315

This section on cladistics will be ended with capital letters – NO SPECIES CAN BE CONSIDERED ANCESTRAL TO ANY OTHER. Why? - NATURES ORDER IS NOT SEQUENTIAL. If the order of organisms is not sequential, then no species can have an empirical standing as an ancestor to any other species. If this is true, then cladistics fails to verify the theories held on organic evolution. Cladistics can only be seen as another form of non-evolutionary classification. **If many evolutionary biologists view Cladism with suspicion and have lost confidence in it as a true thread in the fabric of evolution, then how might this new system of classification be viewed by creationists? Without a doubt they see it as a point of controversy in the evolutionary camp.** I would suggest that evolutionists hand over this new system of typology to creationists. After all, the so-called New Systematics of evolutionists comes dressed up in the garment of the Old Systematics of Linnaeus. **Cladistical studies do a great deal more for creationists in drawing attention to the discontinuous course of nature and the hierarchy of species than they do for the evolutionists who only imagine that they see ancestors in the interpretation of their analysis of cladistics.**

The two methods discussed in this chapter, Biochemistry and Cladistics, do not solve the problem of fossil transitions not being discovered in the geologic strata. The evolutionists' quest will have to continue in their search to find a more feasible answer in response to the question of why there is a lack of intermediates or smooth transitions among species in the fossil record.

CHAPTER 6

THE *ARCHAEOPTERYX* ANSWERS THE QUESTION – "AM I A DINOSAUR?"

(Is the Dinosaur a Connecting Link Between Reptiles and Birds?)

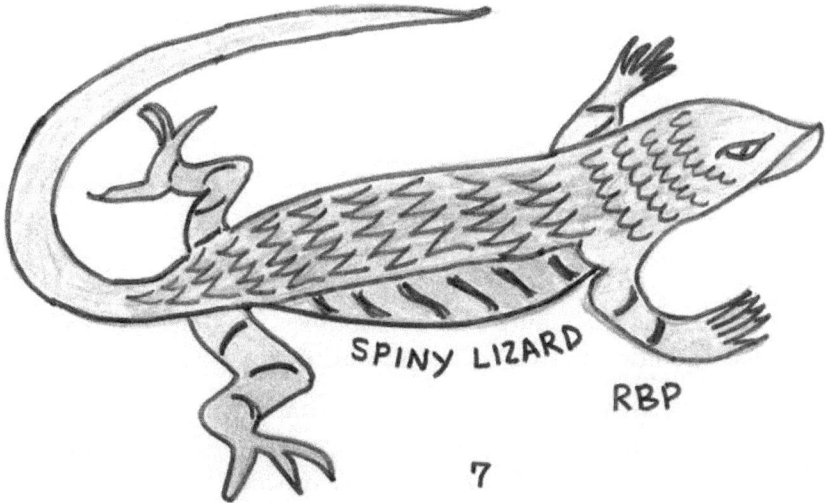

SPINY LIZARD

RBP

7

"Look at the behemoth which I made along with you and which feeds on grass like an ox. What strength he has in his loins, what power in the muscles of his belly! His tail sways like a cedar; the sinews of his thighs are close-knit. His bones are tubes of bronze. He ranks first among the works of God…"

Job 40:15-19

NIV

The Reptiles, according to evolutionists, evolved before dinosaurs in the Geologic Time Scale. The first reptiles appeared in the Carboniferous Period about 345 million years ago. The first dinosaurs arrived about 120 million years later during the Triassic Period. In spite of this fact, the dinosaurs will be our first consideration because of their alleged intermediate status between reptiles and birds.

In *The Descent of Man*, Darwin writes this short sentence:

"We have seen that birds and reptiles were once intimately connected together …"

P528

I suppose Darwin was referring to his words a number of pages back in this same chapter – "Affinities and Genealogy":

"...Prof. Huxley has discovered, and is confirmed by Mr. Cope and others, that the Dinosaurians are in many important characters intermediate between reptiles and certain birds – the birds referred to being the ostrich-tribe (itself a widely diffused remnant of a larger group) and the *Archaeopteryx*, that strange Secondary bird, with a long lizard-like tail."

Ibid. Pp.521-523

Darwin's words coupled with Cope's and Huxley's findings remain a potent force in evolutionary theory even today. I am not familiar with the proper percentages but I do know that many evolutionists hold to the belief that birds are related to dinosaurs. To a creationist, like myself, this might appear radical but to an evolutionist it is only a matter of a slight reorientation of a branch on the phyletic tree.

As Darwin has written, Cope noted characteristics shared by dinosaurs and birds. In fact Edward Drinker Cope (1840-97) and Othneil Charles Marsh (1831-99) were the first to publish papers noting the resemblance of birds to dinosaurs. But it was the English naturalist Thomas Henry Huxley (1825-95) – an authority on bird evolution and Darwin's "bull dog" on the theory of evolution – first noted characteristics shared by dinosaurs and birds.

It is almost comical the way modern evolutionists carry on with the relationship of birds and dinosaurs. The following are a few quotes that should make some evolutionists smile and all creationists laugh:

Robert Bakker is my favorite dinosaurologist. Eighteen years ago I read his book *The Dinosaur Heresies* and have become a student of dinosaurs ever since. [Before my stroke, I had the privilege of lecturing on dinosaurs in various private and public schools and churches. If God gives me the strength, then I will write a third book, which will have to do with questions asked by creationists that followed my lectures. Incidentally, there will be a major question posed by evolutionists]

Robert Bakker is a wonderful artist as well as a picturesque writer. He sketched a large number of drawings for his book and his illustrations show a great deal of imagination and scientific skills. His writing brings technical things down to earth and he never lacks the ability to explain difficult issues. There is by no means a doubt about the nature of his convictions on any issue. The following paragraph gives his position on dinosaur and bird relationships and will give you some idea of his style of writing:

"A truly scientific skeptic would start by assuming neither cold-bloodedness nor warm-bloodedness, and then reevaluate the evidence without prior terminology bias. So long as the Dinosauria remain stuck in the Class Reptilia, this type of analysis is nearly impossible. Let dinosaurs be dinosaurs. Let the Dinosauria stand proudly alone, a Class by itself. They merit it. And let us squarely face the dinosaurness of birds and the birdness of the Dinosauria. When the Canada geese honk their way northward, we can say: 'The dinosaurs are migrating, it must be spring!'"

The Dinosaur Heresies

P.462

Bakker is writing about his proposal for a new classification. He wants to remove dinosaurs from Reptilia, drop the conventual class Aves [birds] and unite birds and dinosaurs under the new class – Dinosauria. I suppose this would throw the birdie people [ornithologists] into a frenzied state of mind. Bakker speaks like a true lover of dinosaurs but I do not belief his classification scheme will ever be accepted in the scientific community.

Alan Charig, Curator of Fossil Amphibians, Reptiles and Birds at the British Museum (Natural History), who is no friend to "birds are dinosaurs" or the new proposal of classification of dinosaurs and birds has this to say:

"Since none of these ideas has yet gained general acceptance – and, indeed, they may never do so – it follows that the suggested reclassifications are hopelessly premature and cannot be taken seriously. They will, in any case, be ignored by most people; after all does it really make sense to call *Tyrannosaurus rex* a bird, or a sparrow a dinosaur? If the latter idea caught on we might well have to change some of our proverbs: 'Dinosaurs of a feather flock together', and 'A dinosaur in the hand is worth two in the bush'. The dawn chorus of the dinosaurs would awaken us early in the morning, we should visit the Dinosaur House at the zoo to see the humming-dinosaurs flitting lightly from flower to flower, and we should round off our excursion by lunching on Dinosaur Marengo in the Members' Restaurant. But, in truth, we need not fear that such absurdities will prevail. Common sense and popular usage are far more powerful influences on language than is scientific pedantry – especially when the latter is based on such insubstantial grounds."

P.140

Getting back to quotes on dinosaur and bird relationships, in spite of what Charig has written, I judge Bakker's above comment to be a classic example of drollness. I don't mean to say that Bakker is not serious regarding his suggestion but he has a clever and humorous way of stating it.

There is more to come! Don Lessem figures not to be outdone by Bakker, his fellow evolutionist. He writes:

"You could say dinosaurs are not gone. You see them flying every day. You eat them when you order McNuggets. Birds are living dinosaurs. As most scientists now think, birds are descended from meat eating dinosaurs. There are many features in the skeletons of meat eating dinosaurs that are shared by birds and no other animals……… …."

"Just which birds are most closely related to dinosaurs we don't know. There are very few fossil birds known from the time of dinosaurs. Most of them were turkey-size or smaller."

Time for Learning Dinosaurs

P.46

You heard it right from the mouth of a dinosaur expert who has written dozens of dinosaur books; each time you go to McDonalds and order McNuggets, think of how tasty your breaded coelurosaurs are! [This comment is for those of you who continue to believe that all birds are "reincarnated" coelurosaurs]

Louis Jacobs writes: "Nevertheless the hypothesis, based on the anatomy of birds …. That all birds, not just *Archaeopteryx,* are coelurosaurs, and that they evolved specifically from an older, as yet unknown ancestor."

Quest for the African Dinosaurs

P.169

Notice that Jacob makes use of the word "specifically." If I believed in evolution, I suppose I would have a "specific" understanding that a life form which evolved, came from some organism that was older than the life form itself. How else could it be? The word "specifically" is superfluous. But when the reader looks for the important feature of the tracing of the ancestral pathway he is disappointed, for the ancestor is "unknown." I have dealt with Jacobs' "so-what hypothesis" at the beginning of chapter four.

Notice also that I have used the word "reincarnated" to express the idea that there are no transitions or ancestral pathways leading from coelurosaurs to modern birds and so the entire alleged phenomenon seems analogous to an incarnation – the coelurosaurs die and millions of years later they appear as living birds which, in actual fact, are coelurosaurs.

I promised you that I would give quotes that will make the evolutionist smile and the creationist laugh out loud. But the following quote is from Christopher McGowan's "sobbing boy" illustration. Only the coldest individual would laugh! I refuse to relate my reaction to this bleeding-heart story. I wonder what your reaction will be! This is the story:

"Every year, during the spring school break, the Royal Ontario Museum organizes an extravaganza of special activities to entertain and enlighten young visitors and their parents. ....

"One year I was told by one of our guards of a distraught little boy who was sobbing his heart out in the dinosaur gallery. When the guard explained what had upset him I dashed off to find the youngster, but by the time I got to the gallery the boy had gone. What a pity, because I think I could have dried his tears. The reason was sad; he had told the guard between sobs, that all the dinosaurs were dead. I would have put a comforting arm around his shoulder and explained that the dinosaurs were really not dead at all – they live on today as birds."

Dinosaurs, Spitfires, & Sea Dragons

P.318

Take heart, Christopher McGowan! Perhaps this little boy read your book years later and was saved from his mental dilemma. Or perhaps on his becoming a man, he may have changed his mind about dinosaurs. He would have found out that *Tyrannosaurus rex* had sharp, six-inch long teeth and if living today, could have swallowed a man whole or have chosen to eat him a little at a time by ripping off large hunks of meat; found out that *Deinonychus* killers, if living today, could have attacked entire crowds of humans and wiped them out by slashing them to death with their vicious and deadly claws.

From another standpoint in viewing past life events, I am happy that McGowan never got to tell the little boy the things that he wanted. If he had comforted the boy with his words, then picture this youth sitting out in his back yard observing a sparrow and trying to imagine his favorite dinosaurs such as *Triceratops* and *Ankylosaurus*. I can almost hear him as he mutters in a soft tone, "Who was that man trying to kid?"

[Pittack's Note: I hesitate in having to mention the following fact but it is essential to our understanding of the theory of the reptile and bird connection. The young boy in the above story could not have imagined *Triceratops* and *Ankylosaurus* for the simple reason that McGowan, should he have caught up with the boy, would have had to explain to him that *not all dinosaurs were living in the bird form.* In fact, the last sentence in McGowan's book preceding his "sobbing boy" illustration is: "There is no question that birds are the direct descendants of theropod dinosaurs (Ibid.P.317). In the beginning of McGowan's *Epilogue* he makes this observation: "Yes, birds are the most delightful creatures. How fitting that they should be the living descendants of the dinosaurs. Making the connection between a songful seed eater and a murderous meat eater requires a certain leap of faith, but the gap narrows considerably if account is taken of *Archaeopteryx*, the earliest bird" (Ibid.P.312). I love the way this evolutionist writes out his thoughts! In one breath McGowan makes his point that birds are the direct descendants of theropod dinosaurs. There is "no question" about the fact of this guarantee. In another breath he urges that such a belief "requires a certain leap of faith."

"How fitting" is this whole idea to McGowan. He should have used the words "How ironic" or "How paradoxical" it is to connect birds with theropod dinosaurs]

So you understand that the young boy could not possibly have thought of *Triceratops* and *Ankylosaurus. These species are DEAD* – the very reason for his sobbing in the fist place. He would have had to think only of theropod dinosaurs as they supposedly lived onward in their LIVING bird form. But what is more – he would have his imagination drawn to species of dinosaurs that fall into the category of "murderous" meat eaters. However, I don't suppose that the museum guard ever planned on giving the boy a detailed explanation. As long as the boy would believe that ALL dinosaurs took their form in birds, this plan of the guard would have had to suffice.

Since McGowan mentions that *Archaeopteryx* was the earliest bird and if taken into account would narrow the gap between dinosaur and bird, let's review this issue seeing that *Arxchaeopteryx* is the main feature of this book. Most evolutionists respect the fact that Richard Owen, back in the eighteen century, was the final authority on comparative anatomy. But they feel because of Owen being a creationist and his ambivalence toward the theory of evolution, he was not willing to mention in his report to the Royal Society the primitive bird (*Archaeopteryx*) had certain reptilian characters. Owen identified this fossil unequivocally as a bird though one of a primitive kind.

But evolutionists, unsatisfied with this identification, turned to the "meticulous" observations of Thomas H. Huxley that *Archaeopteryx* was remarkably similar to a dinosaur. Huxley later assumed that birds had evolved from dinosaurs and this view was upheld until 1920 when Gerhard Heilmann pointed out in his study on avian ancestry that theropod dinosaurs (the coelurosaurs) lacked the furcula (wishbone) present in the *Archaeopteryx*. And so Huxley's theory fell into disfavor until the modern time, John Ostrom came along.

Because some of the new theropods have clavicles and since Ostrom had studied the comparative anatomy of *Archaeopteryx* and theropod dinosaurs (both possessing clavicles), he brought Huxley's theory back into the limelight. While it is true that many evolutionists now accept the hypothesis that dinosaurs gave rise to birds, I suppose just as many evolutionists disagree. Birds' springing from dinosaurs is not a cut and dried issue with all the major camps of evolution. The previous chapter has already discussed cladistic analysis but "The Handy Dinosaur Answer Book" reminds us:

"According to cladistic analysis birds share some 132 characteristics with dinosaurs. Some scientists believe this hard evidence indicates that birds are, indeed, a kind of dinosaur. But many scientists still disagree ...

No, not all paleontologists believe that birds are dinosaurs — or even that birds evolved directly from dinosaurs. Many feel that birds and dinosaurs descended from a common, older ancestor, and developed many superficial similarities over millions of years due to what is called convergent evolution: because both dinosaurs and birds developed body designs for bipedal motion, they eventually started to resemble each other."

However, those evolutionists who are in favor of the dinosaur to bird transformation are never in doubt of their position. They favor the coelurosaurs – those fierce and nasty creatures which are known as the "hollow-tailed reptiles." They were *Compsognathus, Ornitholestes,* and *Coelophysis.* The following quotes demonstrate that coelurosaurs are definitely the IN-GROUP of the dinosaur- bird connection.

Stephen Jay Gould wrote in *The Panda's Thumb* and pages 268-269, "The very close relationship between *Archaeopteryx,* the first bird, and the group of small dinosaurs called coelurosaurs has never been doubted. Thomas Henry Huxley and most nineteenth-century paleontologists advocated a relationship of direct descent and derived birds from dinosaurs."

Adrian J. Desmond states in *The Hot-Blooded Dinosaurs a Revolution in Paleontology* and page 143, "The most likely origin of so many coelurosaurian features in *Archaeopteryx* is by direct inheritance from a small coelurosaurian ancestor ....

There should no longer be any need to have recourse to convergence; the overwhelming evidence now demands a direct blood relationship."

Gregory S. Paul comments in *Predatory Dinosaurs of the World* on page 257, "On the other hand, as the first of the birdlike theropods for which good remains are known, coelophysids are critical to understandings bird origins .... Small terrestrial bipeds such as *Coelophysis* were the real proavians."

Tim Gardom with Angela Milner (Scientific Adviser) wrote in the natural history museum guide *The Book of Dinosaurs* and page 17, "*Archaeopteryx* also appeared in the Jurassic Period. This is the earliest known bird and retains many distinctive dinosaur features, proving that birds actually evolved from small meat-eating dinosaurs."

From the above quotes, we have a pretty good idea that those scientists who believe that birds can be traced back to dinosaurs are very sure of themselves. Of this alleged fact, we have read such phrases "have never been doubted"; "evidence now demands"; "critical to understandings bird origins"; and "proving that birds actually evolved from small meat-eating dinosaurs."

[Pittack's Note: Stephen J. Gould, in his above statement, claims that *Archaeopteryx* and the small coelurosaurian dinosaurs are closely related. Previously, he spent some time on establishing the fact that through a later discovery, coelurosaurs and birds, including *Archaeopteryx,* are very closely related by their furcula "better known to friends of Colonel Sanders as a wishbone" (Ibid. Pp.268-269). Then, he closes his chapter on "The Telltale Wishbone" by urging us to "Think of *Tyrannosaurus*, and thank the old terror as a representative of his group, when you split the wishbone later this month" (Ibid. P.277).

But why did he tell us to think of *Tyrannosaurus* as a representative of its group? Its group was the Tyrannosaurids ("tyrant reptiles") which included *Tyrannosaurus, Daspletosaurus,* and *Albertosaurus.* Gould was referring to the celebration of Thanksgiving when we split the turkey's wishbone. Did he mean that people should think about the wishbone of the *Coelophysis* and be thanking it as a representative of its group? Its group was the Coelurosaurs ("hollow-tailed reptiles") which included *Compsognathus, Ornitholestes,* and *Coelophysis.* This invitation to "think and thank" would have made a lot more sense]

Now is the time when we should consider the question proposed for *Archaeopteryx* – "Am I a dinosaur?"

I have at least one-hundred books in my personal library on dinosaurs. If the juvenile and books for children are included, I have in the neighborhood of well over a hundred. All the *technical books deal with the origin of birds.* One thing I have learned on this subject and have learned it first-hand – the derivation of birds is anything but certain and especially the origin of *Archaeopteryx.*

Gerhard Heilmann after facing a very difficult problem of birdlike dinosaurs not having any collarbones and therefore could not be the ancestors of birds, Heilmann came up with a different theory for bird beginnings. In the 1920s he made the suggestion that if birds did not evolve from dinosaurs, then birds and dinosaurs had evolved from a common ancestor – the predatory reptiles known as "pseudosuchians" ("false crocodiles"). According to Heilmann, two evolutionary pathways made their way up through the Geologic Column from these crocodiles. The one pathway led to birds and the other pathway led to dinosaurs. Instead of the dinosaurs being the "fathers" of birds, they were only distant cousins.

We have already discussed how recent discoveries of some theropod dinosaurs revealed the fact that they possessed collarbones. This re-established the theory of Huxley and Heilmann that dinosaurs were, after all, the ancestors of birds (*Archaeopteryx* was included since one of the seven specimens was discovered with a collarbone). The one evolutionary pathway, once again, was open. But you know how some scientists are!

They hate to give up an old hypothesis. In 1972 some evolutionists took hold of the idea that birds may have originated from the light built Triassic crocodiles. This proposal grew from the study of skulls of living birds and crocodiles. The skulls were compared with each other and then to fossil crocodiles; *Archaeopteryx* and its structures were not even considered in this paleotological study. This was considered a lapse in conventional procedures and few accepted the "bird from crocodile" theory. Larry D. Martin, an ornithologist, accepts a modified version of the pseudosuchian ancestory of birds. This makes birds a sister group to crocodilians. He writes in a newspaper article:

"The theory linking dinosaurs to birds is a pleasant fantasy that some scientists like because it provides a direct entry into the past we otherwise can only guess about. But unless more convincing evidence is uncovered, we must reject it and move on to the next better idea."

["The Barosaurus Is no Five-Story-Tall Canary," *Sunday World-Herald*, Omaha, Nebraska, 19 January 1992, p. B-17]

In 1970 Peter Galton proposed that the "bird-hipped" (ornithischian) dinosaurs were the closest relative of birds. But he later admitted that all known ornithischian dinosaurs were far to specialized to have given rise to birds. So he was forced to come with another idea that was worse than the first: birds arose from some unknown archosaur of the middle Triassic age, which had a bird-like pelvis. Since no creature like the one Galton described has ever been found, he admitted that he was wrong.

Many theories arose in the 1970s concerning the dinosaur to bird issue. 1970: ornithishians to birds. 1972: crocodiles to birds. 1973: theropods to birds.

Other scientists still push the theory that birds evolved from dinosaurs. But many scientists still disagree with those cladists who *believe that birds are a kind of dinosaur* and those who *believe that birds are truly dinosaurs*. We have already read about examples of these particular scientists. But we must ask ourselves the question why so many evolutionists disagree with the affirmative answer to the question posed for *Archaeopteryx* – "Am I a dinosaur?" Why are many scientists claiming this theory of bird origins to be "for the birds"?

The creationist Nicholas Comninellis, M.D. informs us that there are scientists who do not believe birds are dinosaurs or even that birds evolved directly from dinosaurs. He writes "Their rationale is twofold:

Firstly, "Physical distinctions separating birds and reptiles are striking. Reptiles and dinosaurs have anatomy that is, for example, completely incapable of flight."

Secondly, "Analysis of fossils has yet to reveal any creatures with distinctive features intermediate between those of reptiles and birds."

Creative Defense (Evidence against Evolution)

P.164

The twofold rationale of these scientists is here briefly explained ……. …..

Firstly, Body Size Differences: Because of body size differences, it was deemed by scientists as most impossible for dinosaurs to give rise to birds. Carnivorous dinosaurs were ground-dwelling animals with heavy balancing tails and abbreviated forelimbs; the type of bulky body that one would not expect to evolve into a lightweight and streamlined, flying creature.

Secondly, Analysis of Fossils: Through examination of birds and dinosaur skeletons, they appear in many ways to be similar. But there are major differences when it comes to various features of anatomy. The teeth of theropods were curved and serrated. But early birds had peg like teeth, straight and unserrated; dinosaurs, unlike birds, had a major lower jaw joint; the bone girdle of each creature was divergent; birds have a reversed toe for perching while not one dinosaur possessed this character; the ankle bones of birds are unable to demonstrate any homology with those of dinosaurs.

Doctor Comninellis has cited but two examples of the scientific rationale for evolutionists rejecting dinosaurs giving rise to birds. But two examples from other sources will also serve as references ……. …..

The first example: Discrepancy between dinosaur and bird fingers.

Thomas E. Svarney and Patricia Barnes-Svarney write in *The Handy Dinosaur Answer Book* and page 325:

"Dinosaurs developed hand with three digits, or fingers, labeled one, two, and three – corresponding to the thumb, index, and middle fingers of humans. The fourth and fifth digits, corresponding to the ring and little finger of humans, remained as tiny bumps, which have been found only on early dinosaur skeletons. However, as recent studies of embryos have shown, birds developed hands with fingers two, three, and four, corresponding to our thumb and little finger, were lost. Some paleontologists wonder how a bird hand with fingers two, three, and four, could have evolved from a dinosaur hand with fingers one, two, and three. These scientists assert that such an evolution is impossible."

The second example: The Timing Problem; according to evolutionists, there are specimens which provide evidence that birdlike dinosaurs evolved 30 to 80 million years after the *Archaeopteryx lithographica*. If birds descended from birdlike dinosaurs, then it is incongruous to have birds preceding dinosaurs.

The greatest timing problem, in my estimation, is the one that evolves *Archaeopteryx* and *Compsognathus*. *Compsognathus* is the smallest dinosaur fossil found to date with the exception of the "mouse lizard" (This fossil was an actual hatchling of *Coloradisaurus* and when fully grown would have been larger than *Compsognathus*). John H. Ostrom is a strong advocate in promoting the theory that birds evolved from a small coelurosaurian dinosaur similar to *Compsognathus*. He also pointed out the similarities of the coelurosaurs and the *Archaeopteryx* as both maintaining avian characteristics.

It has been mentioned that the Solnhofen quarries of Bavaria, Germany are important to the cause of paleontology because over 600 species have been recovered. The remains of *Compsognathus* and *Archaeopteryx lithographica* were among those fossils discovered. The frequent question has been asked by creationists and rightfully so, *How can the parent be as young as its offspring?* In spite of the similarities that exist between the smallest dinosaur and the first bird, they were *contemporaries* and according to evolutionists were found in the Upper Jurassic and both celebrating their birthday of about 150 million years.

In approaching another problem, some evolutionists see the above similarities as a matter of convergent evolution. But the evolutionist Dr. David Norman writes:

"The question which has to be asked is this: is it reasonable to suppose that all the many similarities really arose as a matter of convergent evolution? The answer would seem to be a definite no. If the similarities been fewer there would always have been an element of doubt, but the probability of so many similarities being produced coincidently must be extremely small. This is not to say that there is no further discussion on this matter of bird origins."

The Prehistoric World of the DINOSAUR

P.175

As a creationist I wonder about the many similarities of the two creatures just as Dr. Norman does. However, I do not believe that *Archaeopteryx* and other birds arose through the coelurosaurs and neither do I believe in convergent evolution which is the assumption that distantly related species have independently evolved similar structures or features. This is not a question of either/or; *either* I believe that birds arose from dinosaurs *or* accept convergent evolution. I believe there is a third option: the appearance of reptilian similarities in the morphology of birds has led me to accept the notion: *Archaeopteryx* was merely an unusual creature and one of the created kinds which has become extinct along with the moas and the dodo.

Perhaps a few creationists view the similarities in a different way. Although I have never come across any creationist who believes *Archaeopteryx* to be a hybrid, I am willing to discuss the issue since it is a possibility that such a belief is presently being held. Is it entirely unreasonable for creationists to view *Archaeopteryx* as a hybrid? For a moment, let us consider the possibility of such rationale. Since the word hybridization is used to allocate both narrow crosses and wide crosses, consequently *Archaeopteryx* could have inherited the characters of a small, bipedal dinosaurs and the most highly developed birds. In other words, the *Archaeopteryx* could be considered a hybrid between individuals that belong to the same large group – the result of the natural crossing of two different types of organisms.

Whether *the one parent is classified as dinosaur or not*, the fact remains; hybridizing members are always bound together by similar morphological characters, have compatible physiological characters, and are always members of the same large group. *Archaeopteryx* could be viewed as a hybrid of the coelurosaur and (?). How else would a creationist explain the remarkable similarities? Nature has her established hierarchies with distinct taxonomic groups. It has its major categories such as families, orders and larger taxa but the Scriptures teach variation and do not say there could have been no crossing between different types of organisms as long as they are members of the same large group.

If there are any special creationists who believe *Archaeopteryx* is a hybrid, they know the most hybridization can do in the matter of change is to give rise to another variety within some already existing kind. On the one hand, they know this is not evolution in the same sense that evolutionists use it in their theory of convergence or the large-scale change called macroevolution – the gradual transformation of one species into another completely different species, similar to the case above – birds evolving from dinosaurs. On the other hand, special creationists know hybridization involves microevolution – which is small-change within some already existing kind, variation within well-defined limits, new varieties produced by crossbreeding and hybridizations.

Chromatin is in the nucleus of a cell and passes on hereditary characteristics from one generation to the next. During certain stages in the cell's life, chromatin is in the form of rods called chromosomes. Genes occur in a linear sequence on the chromosomes and is the fundamental unit governing the transmission of hereditary characteristics. In organisms with sexual reproduction, half the genes come from one parent and half from the other parent. From this point, Harold G. Coffin explains the process called hybridization:

"New changes may arise when two somewhat different organisms cross with each other. This is called hybridization. The offspring may show a blend of the two parents, characteristics of both parents, characteristics more like one parent than the other, or characteristics unique to itself ...

"Because animals and plants which are quite different from each other will not cross, hybridization has definite limitations. Variation in the offspring produced by crossbreeding is limited by the genetic potential of the parents ... .

"It is thus easy to see how hybridization can produce differences, but it is also easy to see that these differences cannot be great. This process does not have the ability to change organisms from one kind into another kind or to produce progressive evolution from simple to complex. The offspring resulting from the crossing of two different parents may be different from the parents but seldom can it be said to be more advanced that the parents."

Earth Story

Pp.101-102

If there are some creationists proposing that *Archaeopteryx* is a hybrid, they may have stepped too far out on the limb. The concept may not be impossible but scientifically unfeasible.

Coffin again writes but this time from a different book, *CREATION – Accident or Design*, P.447:

"Offhand, *Archaeopteryx* does appear to have a remarkable combination of avian and reptilian characteristics. One would be tempted to wonder whether a cross between the classes Reptilia and Aves had actually occurred, but we must admit that hybridization between classes is not seen today .... It therefore seems best, because of the present limited information in *Archaeopteryx*, to consider it an unusual bird, now extinct, that had a position within the diversity of this class that lay nearer the gulf between reptiles and birds than other members of the class."

I do not criticize those creationists who have been tempted to accept the possibility of the hybridization theory but I would suggest that it is a much better idea to stick to the empirical aspects of nature rather than to the philosophical side. When I picture the *Archaeopteryx* in the imaginary scene of taking it home from PETSMART and placing its cage in the special place for viewing, I know it is a unique creature. My common sense tells me that it is a bird but some scientists tell me otherwise. In my mind's eye I keep pointing to the feathers and the evolution-scientists keep pointing to the long, bony tail. I keep talking about the wishbone and the evolution- scientists keep on talking about the claws and teeth. I keep pointing to the primary feathers and the evolution-scientists keep pointing to the jaw.

The question proposed for this exceptional being – "AM I A DINOSAUR?" has been answered in the negative and for the most part by the evolutionists' own words. The rationale for scientists not accepting the theory that birds (including *Archaeopteryx*) ascended from dinosaurs was provided by an adequate number of examples. We must move on to the second question proposed for the *Archaeopteryx* – "AM I A REPTILE?"

CHAPTER 7

THE *ARCHAEOPTERYX* ANSWERS THE QUESTION –

"AM I A REPTILE?"

Is *Archaeopteryx* Any less a Bird Because of its Teeth, Jaws, Claws, Long Tail, Wishbone and Shallow Breastbone?

"Solomon's wisdom was greater than the wisdom of all the men of the east, and greater than all the wisdom of Egypt... ....He spoke three thousand proverbs and his songs numbered a thousand and five. He described plant life, from the cedar of Lebanon to the hyssop that grows out of walls. He also taught about animals and birds, **reptiles** and fish."

1 Kings 4:30, 30-33

NIV

Dinosaur is a compound word coming from the Greek words *deinos* and *sauros* (*dinosauria*). *Deinos* means "terrible" and *sauros* means "lizard." Thus, dinosaur means "terrible lizard." I studied New Testament Greek and related courses in college and the university for a period of seven years. This does not make me a Greek scholar by any means but I do remember that *dunamis* means "power" or "force." In 1866 Alfred Nobel, a Swedish chemist, invented a safe form of nitroglycerin – DYNAMITE (which packs *plenty of power*). The word "Dinosaur" can not be traced back to "dunamis" but it sure sounds like a derivative and that is why some evolutionists chose to call dinosaurs: "Those Dynamite Lizards." But dinosaurs are not really lizards at all since their anatomy is different in five major areas and, unlike lizards; they might have been *warm blooded.* And so, many evolutionists never refer to dinosaurs as lizards. Such scientists chose to call dinosaurs just plain dinosaurs. However, they say that dinosaurs can be traced back to archosaurs and living archosaurs are crocodilians and birds.

To make things more confusing, dinosaurs are in the Reptilia class of animals. Thus, dinosaurs are also called reptiles. But dinosaurs are radically different from living reptiles. Living reptiles are turtles, lizards, snakes, alligators, and crocodiles. Rather than sort out this taxonomic mess, I will stick to what some evolutionists tell us about *Archaeopteryx*. They tell us that *Archaeopteryx* had anatomical characteristics that are similar to a reptile (or lizard since *a lizard is a reptile*). They will also tell us that the *Archaeopteryx* had undeniable bird traits. With these bird and reptile traits, some evolutionists claim that the *Archaeopteryx* is definitely a link between these two classes of animals.

When you think about turtles, you may forget the difference between a turtle and a tortoise, forget the difference between a crocodile and an alligator, forget the difference between a poisonous coral snake and docile, nonvenomous snakes such as scarlet, kingsnake, and milk snakes [These creatures, compared to the coral, have similar markings known as mimicry] but forgetting the difference between a modern day reptile and an ancient dinosaur – I think not. And yet, the evolutionist not only sees a connection but also understands that the *Archaeopteryx* is a connecting link to both the prehistoric reptiles and extinct dinosaurs with the current reptiles. We have already discovered that *Archaeopteryx* could not be traced to dinosaurs. In this chapter, we will learn that *Archaeopteryx* can not be traced to reptiles.

The subtitle of this chapter describes the six major proofs that evolutionists settle upon in their notion that *Archaeopteryx* possessed reptile-like characteristics and therefore is the perfect link between reptiles and birds. Reptiles have teeth, a jaw, claws, a long tail, wishbone and shallow breast bone; so does *Archaeopteryx* – *almost* (It has two jaws). Do these six (More like five since *Archaeopteryx* has *jaws* rather than *a jaw*) reptilian features of anatomy actually serve as proof that *Archaeopteryx* is the link between two classes of animals?

Url Lanham has written a classic book for everyone interested in paleontology in the American West – *The Bone Hunters*. He writes:

"The only way to *prove* by direct evidence that organic evolution has produced the immense variety of living things that now inhabit the earth is to travel back in time to observe the forms of life that existed in the past. So far, the only vehicle known for such an exploration is paleontology." [Emphasis mine]

P.186

Yet, Charles Darwin was forced to rely on indirect evidence for establishing his theory of evolution "since the fossil record seemed to show that novel groups of animals appeared suddenly in the fossil record, without identifiable ancestors." [Ibid.P.186] Darwin mentioned this important weakness of the fossil record but made a prophecy that one day the necessary gaps would be filled with identifiable ancestors serving as missing links and that the fragmentary nature of paleontology would be a thing of the past. Evolutionists believe this prophecy actually came true when *Archaeopteryx* was discovered along with other specimens that make the prophet Darwin a contemporary of his own words. Two years after Darwin's *Origin of Species* of 1859, the discovery of *Archaeopteryx* of the Old World became the "tooth wonder" of paleontology. This first bird had teeth that were alleged to be like reptilian teeth. Some of the "other specimens" were discovered by Marsh as we will soon find out. The "proofs", which had eluded Darwin, were slowly showing up.

Othniel Charles Marsh (1831-99) was an American paleontologist. In the 1800s he was involved in a bitter rivalry with Edward Drinker Cope (1840-97). These two "gentlemen" took part in the "Bone Wars" – the name given to the great rush to find, collect, name, and describe dinosaur fossils from the American West. Marsh's names for dinosaur groups are still used today – sauropods, theropods, and ornithopods. Marsh named some of our most famous dinosaurs. Nineteen genera are to his credit. Marsh's fame was based not only on his ability to find dinosaurs but to discover birds as well. He published in 1880 a volume on the toothed birds of Kansas – *Odontornithes; A monograph on the extinct birds of North America.*

Darwin sent a letter of congratulations to Marsh thanking him for the discovery of toothed birds which helped fill in the gap between birds and reptiles. *Archaeopteryx* had become the "tooth wonder" of the Old world; now it would be complemented by other "tooth wonders" of the New World – *Hesperornis* and *Ichthyornis*. But while Darwin was congratulating Marsh, enemies in the House of Representatives were ridiculing Marsh for spending the taxpayer's money on such a foolish item as bird teeth.

Robert T. Bakker of the present time has his "feathers rustled" as he writes:

"Pure science was a safe target for election-year calumny. Surveys of flooding and coal-mining areas were proper red blooded topics for the Geological Survey, but what earthly use was a thick book about long-dead seagulls that supposedly had teeth?

"Across the Atlantic, in the great university towns and museums, Marsh's monograph enjoyed a totally different response. 'One of the strongest proofs of my theory,' wrote Charles Darwin about the fossil Kansas birds. Thomas Henry Huxley could barely contain his delight: Marsh's birds were a devastatingly effective weapon for beating down the prejudices and half-truths published daily by the anti-evolutionists … ….

"Even European anti-evolutionary scholars (and only a few good ones were left in 1885) regarded Marsh's toothed birds as the long-sought anatomical intermediate between advanced reptiles and modern birds."

The Dinosaur Heresies

Pp.298-301

I once again return to my imaginary caged *Archaeopteryx* from PETSMART and view it while thinking about all the above evidence. Surely it will be more difficult to answer the proposed question put before this bird "of the saintly days of yore" (Poe). I marveled, in my imagination, to hear discourse from *Archaeopteryx* as he plainly squawked, "No-I'm not a reptile!"

But why does this bird refuse to be *pigeon*-holed? Why does it refuse to be removed from its perch and be known as a common reptile? Why does it decline having its bird characteristics compared to a reptile? Since we have already been given insight into Marsh's discoveries of toothed-birds and how Darwin praised Marsh for confirming his theory of evolution; for finding a transition between reptiles and birds, let us consider the first comparison of *Archaeopteryx* with reptiles – TEETH.

The creationist Luther D. Sunderland writes:

"Modern birds do not have teeth but many ancient birds did, particularly those in the Mesozoic. There is no suggestion that these were transitional. The teeth do not show the connection of *Archaeopteryx* with any other animal since every subclass of vertebrates has some with teeth and some without."

Darwin's Enigma

P.75

Sunderland has some good points:

Many ancient birds of the Mesozoic had teeth. Mesozoic means the Era of "Middle Life." Three Periods make up this Era – the Triassic, Jurassic, and Cretaceous. This means there were birds discovered in all strata of the Mesozoic Era of the GEOLOGICAL TIME TABLE. Since the fossil of *Archaeopteryx* (The first bird) was retrieved from the Mesozoic sands of Solnhofen, Germany and Othneil Charles Marsh discovered some of his birds in the Mesozoic chalk deposits of Kansas and they had TEETH; where is it etched in the sands or inscribed in the chalk, we must think *Reptile* and not *Bird*? If for example *Archaeopteryx* had teeth and *Hesperornis* had teeth and they were both birds, then what is the problem? It does not matter that modern birds have no teeth. The *Archaeopteryx* was an ancient bird and he must be matched with other ancient birds – not with reptiles or birds of present time that have no teeth. Some ancient birds did not have teeth and Sunderland reports, "Every subclass of vertebrates has some with teeth and some without."

Therefore, he writes, "The teeth do not show the connection of *Archaeopteryx* with any other animal ...."

The evolutionist Gregory S. Paul writes about the *Archaeopteryx* in his book, *Predatory Dinosaurs of the World* and on page 213:

"In fact, the very conical, unserrated, and big-rooted teeth of *Archaeopteryx* are most like those of marine crocodiles, whales, and the toothed diving bird *Hesperornis*"

Sunderland retorts, "There is no suggestion that these birds were transitional."

He is right. It matters not that he slipped up when writing there is no connection of *Archaeopteryx* with any other animal. Some evolutionists *do connect* the *Archaeopteryx* teeth with the crocodile. But there is not a stitch of proof that there is a "real" connection. I spent a long time in backing up Sunderland's statement of no transitional or intermediary birds with any kind of ancestor in the evolutionary sense and even covered the theory that birds can be traced back to crocodiles. [I would request that my readers refer to chapter three to five for further clarification]

Most statements on TEETH are repetitious in books on creation. But Duane T. Gish's book, *The Amazing Story of Creation* and on page 60, gives a more detailed explanation:

"It is true that modern birds do not have teeth. However, some ancient birds *did* have teeth. By the same token, it is also true many ancient birds *did not* have teeth. The point is no fossils have ever been found that show a gradual disappearance of teeth in birds. They had teeth, or they didn't! This is not surprising, because it is also true of all other vertebrates. Some fishes have teeth, some amphibians have teeth, and some reptiles have teeth. But there are fish, amphibians, and reptiles with no teeth! *Most* mammals have teeth, but some *do not*. The presences or absence of teeth neither confirms nor denies evolution or creation." [The book written by Luther D. Sunderland, *Darwin's Enigma*, is certainly not "repetitious" when it comes to teeth or any other characteristic about *Archaeopteryx*. Sunderland had interviews with five of the most prodigious scientists and top paleontologists at leading natural history museums around the world. They were questioned about their evolutionary beliefs including *Archaeopteryx* and the "reptile to bird concept." Each of the museums involved has extensive fossil collections. The late Sunderland did a service for God and the cause of creationism]

Gish has another good point when he draws our attention to the fact that fossils have never been found showing a gradual disappearance of teeth in birds. Gish merely points out what Darwin's theory of evolution demands numerous transitional links showing the change of characteristics in a particular species. Some evolutionists claim that the transitions are throughout the fossil record but that is a claim that is invalidated by the fossil record itself. In the case of *Archaeopteryx* and all other vertebrate groups of animals; species come on the scene in full perfection, remain substantially unchanged, and then pass away into full perfection. Why then should we expect to observe a charge in teeth either coming or going?

In chapter four, Gregory S. Paul is quoted as saying, "Especially pertinent to our problem (transitional life forms) is how the first birds and bats both appear in the fossil record suddenly and fully developed ....major flying vertebrate groups always appears fully developed so suddenly is very suspicious."

"The presence or absence of teeth neither confirms nor denies evolution or creation," writes Gish. But the word "suspicious" in Paul's statement should remind us that we are *suspiciously close to the creationists' prediction* that life in the beginning would appear fully developed as God originally created the various kinds of animals that seem to fall into the hierarchical scheme of nature. [However, God provided for *variety* in organisms through environmental effects and autogenous variations. Autogenous variations falling under the category of recombination, gene mutations, or chromosome changes]

Let us consider the next comparison – JAW BONES

Archaeopteryx is said to have a lizard's jaw but we will consider the jaws of this first bird for what they are – bird jaws. One of its bird traits are the two jaw bones – the quadrate and quadratojugal not sutured together.

Duane D.Gish compares two papers by evolutionists in order to get their take on the *Archaeopteryx* anatomy of the jaw/jaws and to demonstrate what new methods can do for science research ... ...

1) M.J. Benton states "that the quadrate (the bone in the jaw that articulates with the squamosal of the skull) in *Archaeopteryx* was single-headed, as in reptiles."

Nature: 305: 99 (1983)

2) "Using a newly devised technique, called computed tomography, Haubitz et al. established that the quadrate of the Eichstatt specimen of *Archaeopteryx* was double-headed, and thus similar to the condition of modern birds, rather than the single-headed, as stated by Benton." [Haubitz's report in Paleiobiology 14 (2):206 (1988)]

Evolution: the fossils STILL say NO!

Pp.133-134

In other words, Haubitz found that the Eichstatt specimen of *Archaeopteryx* contained the   maxilla (upper jaw) and mandible (lower jaw). Both jaws moved while eating and this made it SIMILAR TO OTHER BIRDS.

However, *five years earlier*, Benton reported only on the maxilla (upper jaw) as articulating with the middle portion of the skull. During eating, only the maxilla moved. This made the jaw SIMILAR TO MOST REPTILES. Which scientist is correct in his assessment?

In my experience of working in radiology, I have seen what computed tomography can do and so I know what happened in the examination of the *Archaeopteryx* fossil. On the Eichstatt fossil, the CT produced cross-sectional tomographic images. It first scanned the fossil from multiple angles with a narrow x-ray beam, then by calculating a relative attenuation coefficient for the various fossil elements in the section; it displayed the image on a television monitor. This enabled Haubitz to see that the quadrate in *Archaeopteryx* was double-headed.

The above is another example of how further research can change the scientific knowledge of the *Archaeopteryx*. In this case, the characteristic in question is bird-like, not reptile-like.

We will consider the next comparison – CLAWS

Evolutionists point to the claw characteristic as another strong evidence for accepting *Archaeopteryx* as an intermediate form between reptiles and birds. The claws on the feet and feathered forelimbs of the fossil bird are supposedly reptilian and claimed to point to its paternal origin. These points are centered in philosophy but what are the empirical affirmations?

In the first place, fossils of true birds have been found in strata at least as "ancient" as the Solnhofen quarries. If *Archaeopteryx* was a contemporary of modern birds, then it cannot be the ancestor of birds. In February of 1893 O.C. Marsh thought about the birds, *Archaeopteryx*, *Hesperornis*, and *Ichthyornis*. He remarked in the *American Journal of Science*, "the fortunate discovery of these interesting fossils ….does much to break down the old distinction between Birds and Reptiles, which *Archaeopteryx* has so materially diminished" (O.C. Marsh, "On a New Class of Fossil Birds," American Journal of Science and Arts 5, no. 2 (February 1873): 3).

Archaeopteryx looked like a lizard-bird or a bird-lizard and had teeth. *Hesperornis* also had teeth and it was a carnivorous swimmer and *Ichthyornis* could fly. Looking at these three birds "Marsh argued that they were so very different from one another that the evolution of birds must have begun at a much earlier time, perhaps during the Triassic period more than two hundred million years ago, perhaps even at the dawn of the age of dinosaurs."

Mark Jaffe

The Gilded Dinosaur

P.88

The "dawn of the age of dinosaurs" would push bird evolution back another fifty million years (250 m.y.a.). However, these thoughts of Marsh, in 1783, were merely thoughts and not realities. But 112 years later in 1985, the paleontologist Sankar Chatterjee was walking in the scrub plains of the Texas panhandle and he uncovered a few dozen shards of white bones from the red clay and sandstone hillocks.

The reconstruction of the bones became Protoavis – the earliest known bird in prehistory. The remains are alleged to be 225 million years old and 85 million years older than *Archaeopteryx*. I keep making the mistake of referring to *Archaeopteryx* as the first-bird!

I suppose its position has been usurped and it is no longer "numero uno." What does this do to the Geologic Column? This would make the first birds nearly contemporaneous with the first dinosaurs (Just what Marsh thought as he contemplated the bird issue back in 1873). Thus, the possibility of dinosaurs being ancestors to birds was outside the realm of probability (Just what Marsh did not want to consider).

Protoavis is older than *Archaeopteryx* and since this bird is assumed by evolutionists to be closer to Reptilia, it ought to appear more reptile-like but such is not the case. Protoavis is more bird-like than *Archaeopteryx*. Thus, *Archaeopteryx* is not an intermediate between reptiles and birds. And this is not all! Claws on the feet and wings are meaningless to back up the argument for transitional, reptilian characteristics. Then where did these claws come from if they are not a genetic gift from lizards?

Before taking up this question, the main point of all this cannot be over-emphasized. The *Archaeopteryx* can no longer be considered the ancestor of flying birds. Evolutionists must look farther back in time, beyond the world's most beautiful fossil. For the doubters who still maintain that Protoavis is a matter of controversy there are other finds: for example, the bird discovered in western Colorado. James A. Jensen unearthed a fossil in the Dry Mesa of the Morrison formation. This paleontologist from Bringham Young University, in 1977, caused ornithologists (when they first heard the news) to drop their field glasses in disbelief, paleontologists to frantically update their files, and cladists to check their computers. Yes, it is true! - An uncontested bird was found in Lower Jurassic rock at least 60 million years older than *Archaeopteryx*.

Jensen's find invalidates the following argument by Stephen Jay Gould in his book *An Urchin in the Storm*, P.233:

"Rifkin displays equally little comprehension of basic arguments about evolutionary geometry. He thinks that *Archaeopteryx* has been refuted as an intermediate link between reptiles and birds because some true birds have been found in rock of the same age. But evolution is a branching bush, not a ladder. Ancestors survive after descendants branch off. Dogs evolved from wolves, but wolves (though threatened) are hanging tough. And a species of *Australopithecus* lives side by side with its descendant *Homo* for more than a million years in Africa."

Gould's example of the alleged ancestor of *Homo* covers only a million years. Jensen's true bird, as the ancestor of *Archaeopteryx*, covers 60 million years and therefore was more than a mere contemporary. Would this not make Jensen's find a bit problematic for Gould and those contending that *Archaeopteryx* is a missing link between reptiles and birds?

Returning to claws, Duane T. Gish has a hypothetical question for evolutionists which he quickly answers for them:

"Are claws on the wings [of *Archaeopteryx*] evidence of a transition between reptiles and birds? There are at least three birds living today that have *claws* on the wings. The *hoatzin*, a bird living in South America, has claws on its wing when young. This is also true of the *touraco*, a bird living in Africa. The *ostrich* has three claws on its wings, but no one would dare suggest that any of these birds are *intermediates* between *reptiles* and *birds* because they are very much alive and well today." [Emphasis his]

The Amazing Story of Creation

P.60

From the creationist's viewpoint, *Archaeopteryx* is matched with present day birds to avoid the reptilian connection and Gish makes a good point. However, from my standpoint as a creationist, I can see why an evolutionist would make a definite connection between a claw-bird of the present with a claw-bird of the past. Some evolutionists believe the three-fingered hands in the young hoatzin are an atavism. An atavism is the reappearance of a character present in a distant ancestor. Since atavisms are consistent with an organism's evolutionary history, evolutionists "dare" to see the young hoatzin and its connection with the *Archaeopteryx* as a *kind of intermediate*. This would be true since most scientists still believe, in spite of the evidence, *Archaeopteryx* to be a true link between reptiles and birds. These same scientists would see modern birds and toothed birds branching off from the *Archaeopteryx* which in turn branched off from a primitive reptile.

It is difficult to find anything in creationist literature concerning throwbacks, atavisms, or reexpressions by genes so I have to go to my favorite evolutionist and dinosaurologist – Robert T. Bakker. His 14th chapter, ARCHAEOPTERYX PATERNITY SUIT: THE DINOSAUR-BIRD CONNECTION in his outstanding book, *The Dinosaur Heresies*, should be read by every creationist who has even a half-interest in *Archaeopteryx*. He writes in a clear and precise way:

"As birds go, an adult hoatzin exhibits nothing special in the anatomy of its wing. But the young nestling is a genuine evolutionary throwback, an ugly little chick that climbs through the vegetation by grasping with its three-fingered, claw-tipped hands designed to the *Archaeopteryx* blueprint. The hatchling can thus escape predators – snakes and hawks – by using its wing-claws to climb out of its nest and work in to the labyrinth of vines surrounding it …. …. …. ….

"A wealth of evidence supports this theory of reexpressions by genes that have turned off for million of years. Most of it occurs in throwbacks (what nineteenth-century scientists called atavisms). The rare appearance of ancient organs in species that, as a whole, had lost the anatomical features millions of generations earlier."

The Dinosaur Heresies

Pp.316-317

Although "millions of years" cause the creationist to frown, some of the terms are familiar. We know that "blueprint" has something to do with genetics. Also genes "turned off" must be referring to repressor molecules which keep certain genes in the switched-off position in order to keep them from over-producing. The reexpressions, throwbacks, or atavisms must allude to the genes being turned back on after a period of time.

Creationists along with evolutionists realize how important the gene pool is. From this pool, life forms are enabled to develop their characteristics. To the creationist, plants and animals are permitted to vary with limitation. God sets the boundaries of variation (changes in appearance) through environmental and hereditary effects. Limited change in organisms, due to an alteration in gene expressions, is called microevolution.

To the evolutionist, plants and animals develop their variation naturally and without limitation. Nature sets no boundaries in forming species through the environmental pressure of natural selection and mutations (the raw material of evolution). Large-scale change in which one species transforms into another completely different species is called macroevolution. [Although I will not discuss the matter of mutations at this time, I must make this one observation: Evolutionists with their "deep time" philosophy rely too much on mutations. In fact, creationists speak of mutations as the "sole hope" of evolution. On the other hand, creationists with their "short time" philosophy rely too much on recombinations and chromosome changes. Perhaps gene mutations are more important than what creationists believe. Very few books on creationism touch upon the subject of variation among species and if they do, then it is seldom explained, if ever, how variation can take place in such a diminutive time period set by creationism. These books always downplay the possible role of mutation in variation. I think they should take another look!]

Let us get back to the theory of atavism. What is the explanation for claws on the young hoatzin? – Genetics. According to Bakker, hoatzin's ancestors never lost the genetic blueprint for making the clawed fingers style of the *Archaeopteryx*. The genes for making claws were turned off by the physiological switch and then turned on again after a certain portion of time. Evolutionists claim that new advances of genetic research support the theory of reexpression. I would not doubt this claim! Otherwise, how do creationists explain the fossilization of an extinct animal possessing the claw characteristic and the reappearance of the same characteristic in a present living animal? Creation geneticists are aware of reexpression. Also, they are totally aware that genes can be turned off and on again.

For those evolutionists who do not believe in atavism, they would have to belief in re-evolution to explain the claw development. For those creationists who do not accept atavism and do not believe in any kind of evolution, they are left without an explanation for claws appearing in modern species unless they want to conclude that such similarities are coincidental. What have I personally learned through all this? I learned that the baby hoatzin, in adapting to his environment, is given a characteristic from the long ago (not millions of years) for escaping the predators – snakes and hawks; the gene pool and the processes of genetics is mind-boggling and helps reaffirm my commitment to the Creator as He is evidenced through design.; it is an impossibility for me to believe in the fantastic notion that life's gene pools could be turned off for millions of years and turned back on again. This stretches credulity and thrusts reason aside.

"Scratching" out a summary of claws:

*Since there are birds that precede *Archaeopteryx* by millions of years in time and are less reptilian and more bird like, the claws are not an indication that *Archaeopteryx* is an intermediate between reptiles and birds.

*The *Archaeopteryx* is matched with other birds with claws – not with reptiles. What, then, is the problem?

*Although *Archaeopteryx* appears to be transitional to the evolutionist, there are no links between the individual body parts that are fully functional, including the claws. There is no proof that claws demonstrate a blood relationship with reptiles.

*Atavism is no proof of evolution. Nothing new is introduced into the gene pool and thus no new animal is produced. The activation of old genes that were once dormant is no sign of evolution. In fact, atavism is a theory contrary to the belief in re-evolution or double-evolution. It must be admitted; Atavisms are not new differences but simply due to arrangements of genetic elements already in existence.

Let us consider another comparison – REPTILIAN LIKE TAIL

In examining my *Audubon Society Field Guide to North America Reptiles & Amphibians*, I discovered that some reptiles have short tails, many have long tails.

I already knew that but I just wanted to analyze the silhouettes provided in the guide. The fact that *Archaeopteryx* has a long tail: does this make it less of a bird and more of a reptile?

Luther D. Sunderland quotes the evolutionist author Francis Hitching who included a detailed discussion of *Archaeopteryx* in his 1982 book, *The Neck of the Giraffe – Where Darwin Went Wrong*:

"In the embryonic stage, some living birds have more tail vertebrae than *Archaeopteryx*. They later fuse to become an upstanding bone called the pygostyle. The tail bone and feather arrangement on swans are very similar to those of *Archaeopteryx*. One authority claims that there is no basic difference between the ancient and modern forms: the difference lies only in the fact that the caudal vertebrae are greatly prolonged. But this does not make a reptile."

Darwin's Enigma

P.74

I am glad an evolutionist researched this information (Francis Hitching). Creationists are not the only ones to believe the reptilian features of *Archaeopteryx* can be found in birds and that the features do not mean that *Archaeopteryx* is a reptile or half a one or that it is an intermediate between reptiles and birds. Its feather arrangements are more like a swan. Even the tail is more like the swan than it is the reptile. The swan is about 58-60 inches long and has about 25,000 feathers. The *Archaeopteryx* is about the size of a crow. A crow is about one foot to a little over two feet long. Let us take a mean which is 13 inches (12 inches +14 inches =26 inches =13 inches). Feathers would be wildly estimated at 3-4 thousand. The swan (a large water bird) belongs to the genus *Cygnus* and to the family *Anatidae*. The swan belongs also to the Class Aves. *Archaeopteryx* is a perching bird also from the Class Aves.

The bottom line is: *Archaeopteryx* and many lizards have long tails but some lizards have short ones. Also some living birds – in the embryonic state – have more vertebrae in the tail than does *Archaeopteryx*. With these facts in mind it is more reasonable, when trying to determine the lineage of two animals, to compare the *long* tail of the *bird Archaeopteryx* with the *long* tail of a *bird* (in its embryonic state) rather than to compare it with the *long* tail of a *lizard*. That is, it makes more sense to compare a bird with another bird than it does to compare a bird with a lizard and especially when trying to discover the ancestral significance of a certain anatomical characteristic. By the same token, when analyzing a similar ancestral trait, it makes no sense at all to compare the *long tailed bird Archaeopteryx* with a *short tailed reptile*.

Furthermore, the *Archaeopteryx* has the same tail arrangement as the swan. This places the fossil in a relationship with birds rather than with reptiles.

And finally, the feather arrangement on a swan is similar to *Archaeopteryx.* Feathers are the empirical facts of the *Archaeopteryx* and bird connection. This bird association should forever settle the philosophical issue of the reptile and bird connection. The TAIL and feathers have nothing to do with REPTILIA but have everything to do with AVES.

Let us reflect upon other comparisons – The WISHBONE and SHALLOW BREASTBONE

I must confess to never having known about the link of the *Archaeopteryx* wishbone to a primitive reptile until I became familiar with Dr David Norman's book, *The Illustrated Encyclopedia of DINOSAURS*, P.192. Norman writes this thought about *Archaeopteryx* in one sentence:

"The shoulders do however have a distinctly bird-like 'wishbone' or furcula which is supposed to represent the fused collar bones (clavicles) of primitive reptiles."

P.192

Evolutionists attempt to identify every feature of the *Archaeopteryx* as reptilian. This effort to have as many parts of the anatomy of *Archaeopteryx* blend in with a reptile will allegedly establish their theory of common ancestory and set up *Archaeopteryx* as a link between reptiles and birds. Norman uses the word "supposed" which shows the uncertainty of this entire thought. He means that evolutionists presume or expect clavicles of "early" or primitive reptiles to eventually be transformed into the wishbone of the *Archaeopteryx.* A least this is their assumption based on their strong faith in evolution.

Alan Charig, in contrast with Norman, claims with great certainty:

"The most noticeable bird-like features of *Archaeopteryx* are its feathers and its wishbone – two characters which are absolutely typical of birds and have never been found in anything else."[Charig sees no connection of the wishbone with reptiles]

A NEW LOOK AT THE DINOSAURS

P.136

The *Archaeopteryx* is too perfect an animal to show the marks of evolution. I mean by this: there is no evidence that it went through the necessary stages in an alleged evolution. There are no transitions leading up to this bird which has an abrupt appearance in the fossil record. There is no fossil evidence of half-wings, the transitions of teeth, half-claws, or any other organs or features in their intermediary steps. What is more, if this bird is really an intermediate between reptiles and birds, why did it not have primitive feathers instead of completely formed feathers that you see in the present day, advanced birds? *Archaeopteryx* had perching feet, primary feathers, a furcula, wings, all teeth lacking serrations, long external nostrils, two jaw bones, and he flew just like modern birds.

Little doubt that *Archaeopteryx* is a bird and only appears to have reptile characteristics. My readers may recall in chapter six, I suggested that some creationists propose that *Archaeopteryx* was a hybrid. That is how they, no doubt, account for the many likenesses between it and coelurosaurian dinosaurs. I would like to zero in on one more likeness – collar-bones (These bones link the scapulas to the sternum).

Charig continues on with his comments:

"*Archaeopteryx* had a wishbone – the two collar-bones joined together – but all dinosaurs were thought to have lost their collar-bones. A dinosaur without collar-bones could hardly have given rise to an *Archaeopteryx* which still possessed those elements, albeit somewhat modified to form a wish-bone!

"Recently, however, this old idea has been revived and brought up to date. It is now known that a few dinosaurs – coelurosaurian dinosaurs – do have collar-bones ...."

Ibid. P.137

These statements should not surprise those creationists who believe that *Archaeopteryx* was a hybrid. A wishbone and feathers are two "characters which are absolutely typical of birds and have never been found in anything else." Feathers have not been found in any other animal (a contrary belief that is out of step with some evolutionary scientists and must be addressed in chapter 8) but clavicles are a different matter EXCEPT, OF COURSE, when it comes to COELUROSAURIAN DINOSAURS. If certain creationists could not account for *twenty* similar characteristics or anatomical parts existing between a coelurosaur and an *Archaeopteryx*, then how are they able to account for *twenty-one* similar characteristics *except* by purporting that this bird is a HYBRID? Otherwise they would have to accept either the fact of birds evolving from dinosaurs or re-evolution (double evolution); this they would never do.

Those creationists, who believe that *Archaeopteryx* is a hybrid, should be very much aware of the theoretical nature of their concept. Harold G.

Coffin (one of my preferred university instructors of the past) has this advice:

"It is attractive to think that the now extinct bird called the *Archaeopteryx*, with its socketed teeth, clawed wings, and long tail of many vertebrae [was a cross] between bird and reptile ... ... But such suggestions are at present mere speculation. Perhaps information may become available in the future that will warrant further investigation into such possibilities, but until then the idea cannot be given serious attention."

CREATION – Accident or Design?

P.36

Coffin follows this advice with a short dissertation on hybridization:

"Hybridization and progressive evolution are two very different things. Broad crossing could produce what might look like connecting links, but it would not lead to increasing complexity of animal types. Evolution as commonly understood signifies a process that leads from simple to complex through many small transitional steps, until the great variety of living forms, simple and complex, would have been formed. Hybridization of a type suggested here must start with already created major kinds of animals. The virility of plants and animals was greater at creation than now, a fact that may have broadened their capacity for hybridization."

Ibid. P.36

Coffin's above statements do not rule out the possibility of the *Archaeopteryx* as a hybrid but he does lay the matter to rest. Some creationists may claim that the clavicles of *Archaeopteryx* are similar to the coelurosaurian dinosaur but I would suggest the similarities are *coincidental* and that bird characteristics are not entirely unanticipated "considering the broad variation in morphology that is possible in a class" (Coffin).

Duane T. Gish, a creationist, writes:

"With the passage of time, *Archaeopteryx*, in the eyes of some evolutionists, has become more and more 'reptile like'! ... ... some evolutionists today not only assert that this bird is undoubtedly linked to reptiles, but that if clear impressions of feathers had not been found, *Archaeopteryx* would have been classified as a reptile. This is a gross misstatement, since no reptile has avian wings and the many other bird-like features possessed by *Archaeopteryx*."

EVOLUTION; the fossils STILL say NO!

P.141

Perhaps, the HYBRID idea can really be put to rest!

We must leave the wishbone and concentrate on the sternum.

SHALLOW BREAST BONE – Does this comparison draw our attention to lizards or birds?

Everyone interested in nature probably has at least one "Golden Guide" book. I have a dozen or more because once having read one, more has to be purchased. They are easy to carry around (3 inches by 6 inches); easy to read and memorize (each book carries short *sections*); each one is packed with a wealth of information. *A Golden Guide Dinosaurs* written by Eugene S. Gaffney, Ph.D. and beautifully illustrated by John D. Dawson claims:

"Although *Archaeopteryx* had feathers, could it fly as well as modern birds? Probably not. Modern birds have a greatly expanded keel, or breastbone, to which large flight muscles are attached. *Archaeopteryx* did not have this breastbone and so probably did not have large flight muscles. But it could probably fly or glide to some extent."

P.90

A Golden Guide Dinosaurs was written in 1990. Had this book been written but three years later, the text would have been written much differently. We will soon discover why the difference, after running through a short history of *Archaeopteryx* finds:

The *first Archaeopteryx* was discovered in 1861, Germany, and is called the "London specimen." In 1877, the *second* specimen was found also in Germany. It had a crushed head but nicely preserved body. It was called the "Berlin specimen." The *third* bird was found in 1958 by Eduard Opitsch and as might be guessed where – Germany. Opitsch died in 1992 and the fossil died with him. The fossil disappeared and its whereabouts remains a mystery.

The *fourth* fossil was discovered under some unusual circumstances: John Ostrom was studying pterosaur flight and went to the Teyler Museum in Haarlem, Netherlands. Ostrom entered the gallery by way of invitation from the curator, picked up the pterosaur specimen that had been under glass, saw that the tag identified the two slabs were from Solnhofen and immediately knew that he was *not looking at a pterosaur*. He tipped the limestone slabs up to the light and saw the feather impressions of the bird that would later be identified as the forth *Archaeopteryx*. The specimen had been wrongly identified as a pterosaur for more than 100 years. Ostrom had to sit down as he took notes; recovering from his shock. In time, the professor had made famous the Teyler museum. And so the forth *Archaeopteryx* specimen found in Germany in 1855 had to await its true identity for 115 years.

The *fifth* specimen was discovered in 1951 and adopted into the family of *Archaeopteryx lithographica* as the first juvenile. It has the most well-preserved head and measures about two-thirds the size of the other specimens. There is a debate going on over this fossil – is it a new genus or a juvenile? The *sixth* specimen was found in the 1960s near Eichstatt, Germany. It was first identified as a *Compsognathus* in the preparation lab but after discovering the arms were too long for its body size and seeing the feather impressions, scientists recognized the fossil as the sixth *Archaeopteryx* to be found.

So far the first sixth discoveries of *Archaeopteryx* fit the context of *A Golden Guide Dinosaur*, three years later the *Guide* information would be partially obsolete – for in April of 1993 the seventh fossil was discovered and became known as the Solnhofen-Aktien-Verein specimen. What was new about this find? – It was described in 1994 as a *new* species: *Archaeopteryx bavarica* and it has, with the feather impressions, *a small ossified sternum*.

If the information could be rewritten, it would have been written (with the newly discovered information) something like this: "Although *Archaeopteryx* had feathers, could it fly as well as modern birds? Most likely yes – it could fly. Modern birds have a greatly expanded keel, or breastbone, to which large flight muscles are attached. This seventh Solnhofen *Archaeopteryx* was the only fossil discovered with a sternum and although it is smaller than most birds and had smaller flight muscles, it was as excellent flyer."

We have seen the initial discoveries of the first six fossils indicated no evidence of a bony sternum. Paleontologists were able to come to no other conclusion than; *Archaeopteryx* was either a poor flyer or could not fly at all. The seventh discovery in April 1993 changed the scientific reasoning, with reference to the bird's flying abilities, from uncertainty to no doubt at all. *Archaeopteryx*, with its wing and feather pattern of Primaries and Secondaries coupled with its bony sternum, could have mastered the sky with the grace, skills, and flight dynamics of any modern bird. Apparently, because of the delicate nature of the anatomy and due to the smallness of the sternum; the sternum was not seen in the first six discoveries of the Archaeopteryx.

Luther Sunderland quotes Francis Hitching who made a detailed discussion of the *Archaeopteryx* characteristics including its shallow breastbone:

"Various modern flying birds such as the hoatzin have similarly shallow breastbones, and this does not disqualify them from being classified as birds ... ...

"Recent examination of *Archaeopteryx*'s feathers has shown that they are the same as those of modern birds that are excellent fliers. Dr. Ostrom says that there is no question that they are the same as the feathers of modern birds. They are asymmetrical with a center shaft and parallel barbs like those of today's flying birds."

Darwin's Enigma

P.75

Archaeopteryx can in no way be considered a reptile. A small bony-sternum does not automatically disqualify it from being classified as a bird. Before April of 1993, evolutionists did all they could to project *Archaeopteryx* back into time and to link it with primitive reptiles of the Permian Period. But the sternum plus wing and feathers lifted this bird (geologically speaking) from the Late Paleozoic Era to the Late Jurassic of the Mesozoic Era. In other words the *Archaeopteryx* has winged its way (metaphorically speaking) upward and separated itself from retiles by millions of years. Under every GEOLOGIC TIME TABLE the *Archaeopteryx* is listed as a part of the *Main Events of Characteristic Life* in the following manner:

"First (reptilian) bird."

This is a false characterization. Remove (reptilian) because we have already indicated that the main characters can not be associated with or compared to reptiles. Remove "First" since from our studies we know this is not true. There were birds 75 million and 225 million years older than *Archaeopteryx* (according to evolutionary time scales).

So far in this book we have endeavored to bring out certain facts:

- The discovery of *Archaeopteryx*.

- The *Archaeopteryx* fossils are genuine.

- *Archaeopteryx* is not the "perfect" missing link between reptiles and birds.

- Creationists are not unreasonable in asking for fossil transitions.

- Biochemistry and Cladistic Analysis do not solve the problem of missing links.

- *Archaeopteryx* is not a dinosaur nor is it a link between reptiles and birds.

- *Archaeopteryx* is not a reptile and no less a bird in spite of its seemingly reptilian characters.

With these facts under our belt, let us go to the next step in identifying the true characteristics of *Archaeopteryx*. So far, we have been introduced to one theory after another by evolutionists who have sent us through a labyrinth of contradictions and philosophical speculations in their attempt to arrive at a viable viewpoint of *Archaeopteryx*. This is the result of rejecting the imperial facts of nature and depending upon human wisdom to read nature correctly. Many scientists deny the existence of God and turn away from the truths revealed in the book of Genesis. Without the guidelines of the Creator, man has a tendency to wander around in the darkness of his own reasoning. Thus harmony among such scientists seems hopelessly beyond repair. At one time it was the consensus of scientists that debate and difference of opinions amongst themselves was a good thing – this is what science is all about; the only sure way to arrive at truth. However, they seemed to have changed their opinion – harmony is the preferred goal. Holding an International *Archaeopteryx* Conference and signing a general credo of consensual harmony was done "in order to forestall possible misuse by creationists of apparent discord among scientists." But this is nothing other than an attempt to cover up the discord that is much more than merely "apparent."

We will continue with our quest to find the answers to the identity of this fascinating and intriguing creature discovered in the greensand of Solnhofen.

Chapter 8

THE *ARCHAEOPTERYX* ANSWERS THE QUESTION –"AM I A BIRD" (PART I)

(Feather Evolution – Possible or Impossible?)

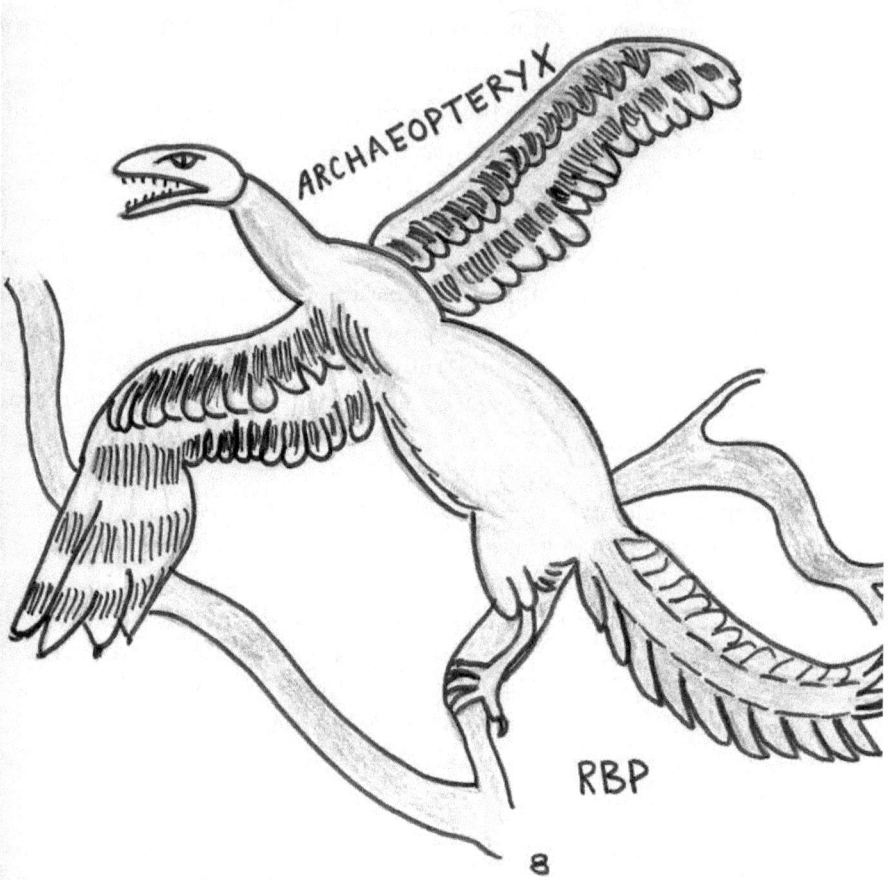

"Do not revile the king even in your thoughts, or curse the rich in your bedroom, because a bird of the air may carry your words, and a bird on the wing may report what you say."

Ecclesiastes 10:20

NIV

We are now ready to ask the proposed question for the *Archaeopteryx* – "Am I a Bird?" That is, "do feathers make me a full-pledged bird in my own right?" The two main bird traits of *Archaeopteryx* are *wishbone and feathers*. We have reviewed the wishbone to some extent; the feathers will be examined to a fuller extent. But before we adorn dinosaurs with feathers where feathers do not belong and write about birds where feathers do belong, feather-evolution would seem to be the logical starting point.

On page 53 of *A Field Guide To Dinosaurs* and written by David Lambert with numerous consultants and advisors, the claim is made that feathers may have evolved from faulty reptile scales that split as they grew. A diagram appears in four stages:

The first stage shows a scale that has two splits at the base.

The second stage shows most of the body and apex separated from the two splits with a third split at the base of the body.

Stage three shows the scale frayed into four parts.

The final stage is a form of "scientific" magic; a fairy tale, if you please! – The faulty scale has been transformed into the most perfect, beautiful, and fully developed feather (vane) that could possibly be imagined. It even has *afterfeathers* at the base of the calamus (quill).This is almost as dubious as Darwin's diagram of the human eye.

At the time of this writing, I am seventy-one years of age. As I go through life, evolution is becoming to me less and less a plausible theory and more to the point of being ridiculous, improbable, and far-fetched, especially when I contemplate the scale-to-feather evolution. In spite of the obscure, scientific terminology and intellectual halo that surrounds the doctrine of evolution, the shear weight of the multitude of theories springing up and out of it, the large number of disciples who study it and the apostles who advocate it, still I think of evolution as no greater than a medieval belief which should have become extinct long ago. My mind will never accept the concept of a one-celled molecule being transformed, in deep time, to multi-celled *Homo sapiens* with a brain and nervous system, bone and muscle system, a circulatory system; all this derived from millions of incredible transformations. Evolutionary explanations for this process are not only *tragically inane* but *extremely superficial* in spite of the cerebral verbiage used to camouflage its shallowness. But I shall, nevertheless, continue on with "feather-evolution."

James F. Coppedge, a research scientist in molecular biology, perceives feather evolution not to be "tragic" (as I do) but rather "amusing" (It is certainly not the "ha, ha" amusing but rather the "pathetic" amusing). He also points out the chances of not only producing a feather but an *entire bird* as well:

"When one understands the complex precision of a feather's design, it is amusing to consider evolutionary claims that feathers evolved from reptile scales!

"To obtain a bird by chance mutations, one would have to overcome the odds against feathers, computed along with the improbability of occurrence of all other marvelous abilities birds would need in order to operate successfully at all. The size of the resulting odds would be a figure that would more than fill the cosmos ..."

Evolution: Possible or Impossible

P.216

Some evolutionists are candid when it comes to the origin of feathers. For example when Luther Sunderland questioned Dr. Donald Fisher (state paleontologist at the New York State natural history museum) "Do you know anything about how feathers might have arisen from scales?" He replied, "None whatsoever" (*Darwin's Enigma*, P.71).

Fisher does not say that feathers did not arise from scales. But rather, he does not know *how* they did. But other evolutionists state with such a display of assurance that either they think they know or are bluffing:

"One feature that living birds have that no other group of organisms has is feathers. Feathers are a body covering, evolutionarily derived from reptilian scales."

Louis Jacobs

Quest for the African Dinosaurs, P165

"....feathers (which, by long-standing professional consensus and clear factual documentation, evolved from reptilian scales) ..."

Steven J. Gould

I Have Landed, P.327

"Feathers are modified reptilian scales," Gould chimes in again.

Bully for Brontosaurus, P.145

Whether feathers have been "derived," "evolved," or "modified" from scales, it is all the same to evolutionists. In this case: "Professional consensus" is a long way from "clear factual documentation."

When Gould writes that there is "documentation" for scale-to-feather evolution – I wonder how many of his readers truly believe his claim! "Factual documentation" exists nowhere on this earth since it defies all reason and logic and is only meant to deceive the gullible public.

P.F.A. Maderson refutes the above assertions by stating his own proposal. He is stating the real truth of the matter of feather evolution:

"I emphasize that this model [figure of a scale transforming into a feather] only attempts to explain how an archosaurian scale might have given rise to a proto-feather. The end product as shown in figure 1d resembles a feather in the usual sense of the word only that it is a highly specialized keratinous integumentary appendage. We cannot offer any plausible explanation for the origin of the unique shaft, barbs, and barbules without which modern feathers would have neither aerodynamic nor insulatory function."

[Maderson quote is found in *EVOLUTION: the FOSSILS still say NO!* by Duane T. Gish, P.136. P.F.A. Maderson's original paper is in "The American Naturalist" 146:427, 1972]

[Integumentary means "having an outer covering or coat, such as the skin of an animal or the coat of a seed. Maderson's entire proposal silences the false statement of "factual documentation"]

Joel Cracraft, an associate professor at the University of Illinois Medical Center has a special interest in functional morphology of birds. Speaking of the feathers of birds and the hair of mammals, he says: "Both these characters are interpreted by biologists to be derivatives of scales."

Gish responds to this remark by writing:

"This is pure myth, empty rhetoric, no matter how imaginative such stories may be. What are scales? They are thin, flat, overlapping horny plates. This horny epidermal covering in reptiles is shed periodically. Feathers, on the other hand, are incredibly complex structures, an engineering marvel; the flight feathers being precisely designed for aerodynamic function …. Feathers are fundamentally different structures than scales, arising from different layers of skin. Scales are merely folds in the epidermis, while feathers and hairs develop from follicles. The development of a feather is an engineering marvel …. …. …… …..

"Here evolutionists like Cracraft are clutching at straws – or should we say feathers?"

Ibid. P.322-323

Gish's clear factual definition of scales contrasted with feathers is called *true science*.

Pro-avis is a hypothetical bird surrounded by numerous theories which have sprung from the best theorists that evolution has to offer. One such theorist was Gerhard Heilman who wrote *The Origin of Birds*. His book advocated the *aboreal theory* which we shall study in our next chapter – "The Flight Capabilities of Archaeopteryx." [The word "aboreal" means walking and running along branches and living in trees] However, prior to Heilman's pro-avis being able to climb trees and live in them, have its gliding transformed into powered flight, it had to be a terrestrial runner – a REPTILE.

Michael Denton evaluates Heilman's scheme as highly speculative. He writes:

"Overall, Heilman …. Attempts no rigorous mathematical aerodynamics approach, which would give estimates of wing area, body weight, and lift at the various stages to show that his "frayed scaled" aerofoil would work and the transition to gliding, and from gliding to powered flight was at least feasible. Indeed, Heilman's imaginary reconstruction of pro-avis … leaves one with the distinct feeling that its wing/weight ratio would be insufficient even for gliding, let alone powered flight."

Evolution: A Theory in Crisis

P.205

Such is the power of evolution! It can paint a vivid picture on the canvass of a man's thought patterns with the oils of imagination and accompanied by the allusion that such a painting is the product of pure science. There is no need for attempting "mathematical aerodynamics" when calculating the results of the frayed scales theory – a brilliant scientist has devised it and that, apart from anything else, should meet evolutionary demands. A deplorable number of the public have regarded the white lab coat of the scientist to become a status symbol of purity; more like the apparel of a high priest who is allegedly free from error and whose every word is a paradigm of truth.

When attending the natural history museums, I never cease to be aware that the specimens on display sometimes are acquired only at great risk and inconvenience to paleontologists and other scientists. I for one appreciate their work and have realized even since my childhood, when viewing reconstructed or restored animals, these specimens prepared for my pleasure and scientific interest did not position themselves magically on their pedestals or have a chance arrangement in their display cases – they had to be the result of many hard working men and women.

It is easy for a creationist to sit back in his easy chair, sipping his cup of tea, and reading the Bible while paleontologists are out in the field, sipping from their cans of beer, and thinking over the scientific significance of some specimen. But while I admire the creationist, I do not disrespect the evolutionist; while I am a friend to the creationist, I am not a stranger to the evolutionist. However, when it comes to the doctrine of creationism versus the dogmas of evolutionism, for me, there is no contest: I am a devoted friend of Creationism but a deadly enemy of Evolutionism. I appreciate the work that scientists perform but protest the lies that support the damnable heresy of evolution which exists in our present society.

But do not think for one moment that I believe every evolutionist to be godless and void of sincerity. There were moments, throughout history, God needed to talk with evolutionary scientists and to let them know that they were included as recipients of His proclamation for the world and just as responsible for their beliefs and actions as are creationists. God spoke through Noah: scientists stood before the ark as Noah spoke with them about the fixed laws of nature – how they do not operate outside the will of God nor are they above the purposes of God. He spoke through Paul who lectured to scientists on Mars hill. Scientists were telling the people of Athens to regulate their lives in accordance with nature but God wanted people to know that it was He who made all nature – the entire world: not an UNKNOWN god. God wanted them to know He was not fashioned (like their idols) from gold, silver, or stone and He will one day judge the world; that He commands people everywhere to repent – not just scientists but all people everywhere and in every place. And what was God's message to the people of Athens? – The same as it is to today: worship God for "we are his offspring" and the message of the brotherhood of man, for "from one man [Adam] He made every nation of men, that they should inhabit the whole earth." God used young Timothy to proclaim "the grace of God that brings salvation has appeared to *all men*." Redemption is for *all individuals* urging them to bring their lives under godly control and to eschew wickedness.

The message of God is not intended for creationists alone but for evolutionists as well. And the message is for *all those individuals outside the scope of these two philosophies.* To assume that all mankind can be categorized under these TWO CAMPS is to assume too much but not only that: such a supposition would be a misunderstanding of the kinds of revelation available to men. For example, I have in mind those people who have lived in remote regions of the world and who have had no access to such information as evolution and creation as opposing forces. God's message including His grace, above all things, will have to reach certain people by whatsoever are His ordained means.

I am not sorry for the lengthy excursion. But we must return to pro-avis and frayed scales. I like the description of frayed scales turning into feathers by Hank Hanegraaff in his book, *The Farce of Evolution* and pages 38-39. I like the depiction because Hanegraaff brings out the true worth of such an unwise belief:

"The pro-avis fairy tale, like any good fairy tale, begins long, long ago with little pro-avises running around on two legs while they playfully caught insects in their scaly little hands. One fateful day an ugly little pro-avis we'll call Mike was born. Unlike his brothers and sisters, little Mickey had frayed scales on both of his hands. Because of little Mickey's imperfections, no one wanted to play with him. Sadly, he has to run around by himself trying to catch insects. Suddenly, little Mickey discovered something miraculous. Insects stuck like magic to his frayed scales. The more he caught, the better he ate. The better he ate, the faster he ran. The faster he ran, the more his scales frayed. In time little Mickey's ugly scales were transformed into beautiful flying feathers. Soon little Mickey was able to catch insects that would normally have been beyond his reach. It wasn't long before all the little pro-avises wanted to be just like Mike. Their scales were transformed into fantastic flying feathers. And they lived happily ever after ……. …."

"In fairy tales, frayed scales turn into feathers, and frogs turn into princes. In evolution all you have to do is add millions of years and little pro-avises turn into beautiful flying birds."

The most fantastic part of this fairy tale is not only the theory of how Mike developed feathers but *how* other little pro-avises became like Mike. It seems to me that Mike had the first stages of feather development happen fortuitously due to environmental effects. How then was each stage of the feather transition from frayed-scales in Mike's reptilian state passed on to his offspring to eventually become perfect feathers in little Mikes' ultimate bird state? How could a chance happenstance – taking place in the environment – become a part of autogenous variations? The environment cannot produce genitival changes for hereditary traits passed on to offspring. The variations, occurring in Mike's offspring, must follow inheritance laws which are in accordance to Mendel's principles. Feathers for wings and wings themselves must be in the original gene pool.

What in the world did a running reptile with frayed scales have to do with the hereditary process of wing development in birds? As a creationist, I understand adaptation which is the mutual fitness of a life form to its environment. However, I do not accept the unsound principle that the physiological state of the genes is influenced by environmental factors. This is too much like "Lamarckism" – the belief that the characteristics of parent animals have acquired from adaptation to their surroundings, can be passed on to their descendants. Science has already proved Lamark's hypothesis of "acquired characteristics" to be untrue. Such characteristics are not inherited.

The way that evolutionists believe concerning natural selection is almost as bad as Lamark's hypothesis of "acquired characteristics." As a creationist, I understand natural selection can select certain gene combinations. Darwinism claims that natural selection is the principal mechanism of modification; coupled with deep time, lower forms of life can be transformed into higher forms of life. James F. Coppedge makes an excellent point:

"Natural Selection Cannot Select What Is Not There! A man cannot select from the shelves of a store what the store does not carry. Before evolution can work, there must be varieties from which to select. The variations, moreover, must offer other improvements which involve surviving or producing offspring. The improvements must also eventually lead to different kinds of animals or plants. Otherwise there is no evolution. But alas, there is difficulty finding a source for new material with such a capability."

Evolution: Possible or Impossible?

P.82

The main key of natural selection that opens up the door to evolutionary change is, supposedly, the ultimate source of all new genetic material – MUTATIONS. However, evolutionists have failed miserably in their efforts to prove that mutations are able to produce large-scale changes where one species transforms into another completely different species. But it is also true: the crossbill (a twisted *mutant* bill) is a successful bird able to separate the "sporophylls of pine cones and extracting the seeds"

(Marsh). There are other examples of plant and animal forms which attest to the role of mutation in producing variety. Mutations are not enough to change the entire genetic code of any particular animal. However, mutation is a change in a gene which can produce a certain amount of variety.

A few changes in the gene expressions of a certain kind of animal will not carry it (even after millions of years) to the pinnacle of macroevolution and become another species. Variation can be carried out in life forms to the point of microevolution – limited variation – but never of the quality which will convert one kind of organism into another kind. For example, the bird kind has been genetically coded to carry out its role in the perpetuation of birds and birds only (the crossbill is still a bird in spite of its mutant bill). Variation remains within kinds and is consistent with the "oft-repeated law" in the first chapter of Genesis: "after their kind."

The claim that feathers arose from faulty reptilian scales is hardly likely and is more like impossible. The fairy tale of frayed reptilian scales transformed into feathers is not physically possible. Feathers are fundamentally different structures than scales. Besides more is evolved than merely the feather or vane – each feather has a unique shaft, barbs, and barbules. With the absence of these features, birds would have neither aerodynamic nor insulatory function.

Pro-avis is a hypothetical bird surrounded by a host of theories with not one iota of empirical evidence to back it up. Also, the theories are against the dictates of true science. Even if the fairy tale of the frayed reptilian scale were true, there was no way for genetics to pass such a trait to offspring. Lamarck's hypothesis of "acquired characteristics" has been rejected even by evolutionary scientists. Having uncovered the irrationality of feather-evolution for my readers, I am now ready to move on to the declaration that dinosaurs laid first claim to feathers.

CHAPTER 9
CHINA AND THE DINOSAUR-BIRD CONNECTION
(Do Feathered Dinosaurs Truly Exist?)

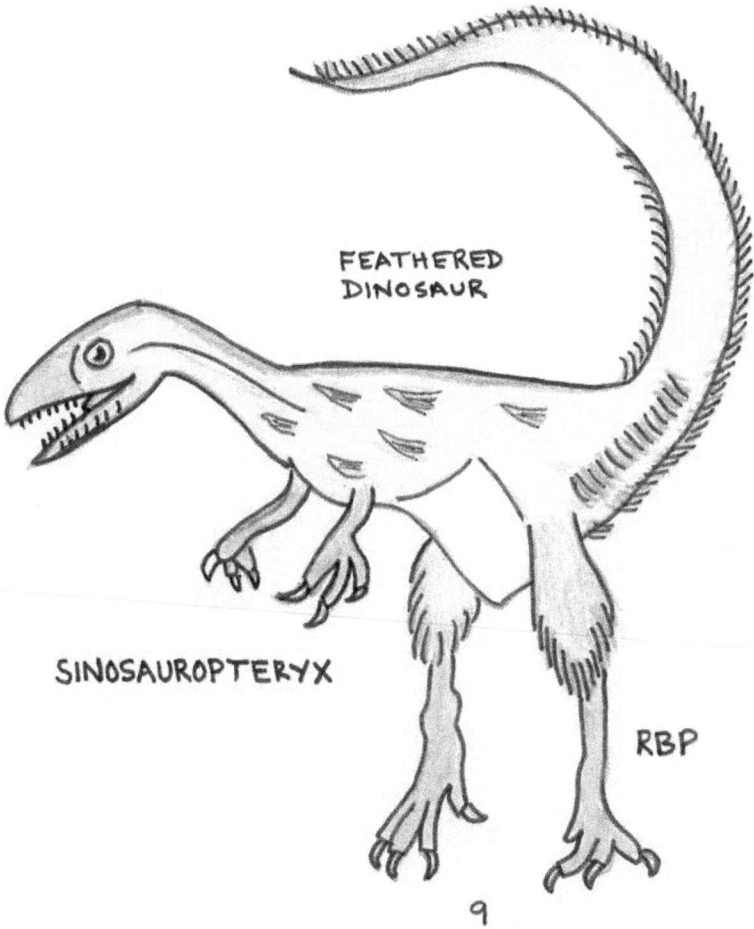

FEATHERED DINOSAUR

SINOSAUROPTERYX

RBP

9

"The trees of the LORD are well watered, the cedars of Lebanon that he planted. There the birds make their nests; the stork has its home in the pine trees. The high mountains belong to the goats; the crags are a refuge to the coneys. The moon marks off the seasons, and the sun knows when to go down. You bring darkness; it becomes night, and all the beasts of the forest prowl."

Psalm 104:16-20

NIV

Before uniting and adorning *Archaeopteryx* with feathers and making it into a full-pledged bird, we need to address this issue of "feathered dinosaurs." We are called upon to revive the old issue of birds – *Archaeopteryx* included – being descendants of dinosaurs. The reason for waiting until now, feathers is a main feature in this book. In that *Archaeopteryx* is definitely a bird and because the wishbone and feathers are the main traits of birds, I thought it to be more logical to save and discuss feathered dinosaurs for this point in time.

Dom Lessem, who publishes beautiful books on dinosaurs, took off in a whirlwind of wild writing in describing feathered dinosaurs. He appeared excited and "keyed up" to get his words quickly into print. Naturally he would be, since his description of the finds of feathered dinosaurs in China *seemed* to confirm everything about what evolutionists had reported in their prestigious papers about dinosaurs being the predecessors of *Arxchaeopteryx* and other birds. First of all, he reports that Microraptor is the smallest dinosaur and no bigger than a crow, currently regarded as the closest relative of modern birds of any known dinosaur. He says, "Even more remarkable is that it had feathers on its four limbs." The following is the main point of his excitement:

"Recently, several dinosaurs with feathers have been found in China. But these feathers were not designed for flight, as bird feathers are. Some were more like needles while others were more like combs. Only birds have feathers that spread in two directions out from a central stem.

"With this feather design, birds can do something no dinosaurs could do. They can fly!"

Time for Learning DINOSAURS

Pp.46-47

Lessem continues in his book:

"A quarry in Liaoning, in northeastern China, is where many remarkable feathered dinosaurs have recently been discovered. Farmers made the first discoveries."

Ibid.P.47

The last I want to do is confuse my readers. Rather than cite other opinions about feathered dinosaurs, I think it's very important to comment on Lessem's remarks before going any further. He is reporting too much in such a few short paragraphs and I am going to slow him down with a few interruptions. First, he claims that feathered dinosaurs have been found in China. The *feather issue* is a point of strong contention as we shall now discover.

In the *National Geographic*, July, 1998, there is an article describing the China discoveries. This expose was written by Jennifer Ackerman and entitled "Dinosaurs Take Wing." The write up mainly had to do with birds descending from dinosaurs and, of course, feathered dinosaurs made the theory more convincing. References were made to the feather issue and the strong controversy that existed then and exists at this present moment in time. For examples:

"News of the Chinese finds has fired up hot debate over the twigs and branches of the avian family tree. Most scientists place *Sinosauropteryx* squarely on an early branch linking dinosaurs and birds. But a few others still doubt the dinosaur – bird tie, holding that the avian clan evolved from some earlier reptile, long before *Sinosauropteryx* feasted on his last meal.

"That a small dinosaur with a hint of kinship to modern birds would ruffle feathers is hardly surprising. The ancestory of birds has aroused much passionate debate as any puzzle in evolution, except perhaps the origin of life itself and the beginnings of our own tribe."

[*National Geographic*, VOL.194, NO.1, Pp.77, 84 (July 1998)]

Ackerman calls our attention to the fact that the dinosaur – bird debate is superceded only by how life originated and how humans evolved from ape ancestors. Ackerman should have added a fourth debate. I have a book in my library entitled *The First Chimpanzee* and on the inside flap it reads, "Recent scientific research has established that Charles Darwin was wrong: humanity is not descended from the apes! .... In other words, apes are descended from humans or protohumans." [Pittack: Some evolutionists never cease to amaze me!]

Let us return to the dinosaur-bird debate. It should become clear to us, as we advance along in this study, why these remarkable fossils fell from paleotological grace and why the winged-dinosaurs finds are still up in the air (pardon the pun!) – Not to mention why the number of skeptics have grown in number.

The paleontologists, who first studied the fossils, noticed the various "signatures" of nature. They noticed along the back of

Sinosauropteryx, from neck to tail, ran a thin, dark ridge of fibrous lines. What could these strange fibers mean? They also noticed hair like structures on *Beipiasaurus* and *Sinornithosarus* there were short fibers. And what did these detectives deduce from these diminutive markings? With such mere clues they were able to establish a series of scientific facts (?). For example:

*Earliest examples of bird feathers.

*Origin of birds and avian flight.

*The placement of *Sinosauropteryx squarely* (Ackerman in *National Geographic,* p.77) on an early branch linking dinosaurs and birds. ["Squarely" in this case must mean the paleontologists did so with a great air of authority. It was as though such a taxonomical arrangement of this dinosaur was irrefutable and "here to stay."]

*Birds evolved from small carnivorous ground-dwelling dinosaurs, etc.

Before I allow Dr. Nicholas Comninellis to "let the air out of the bag" on the strange markings (the thin, dark ridge of fibrous lines, etc.); a quote from *The Nature Companions Rocks, Fossils and Dinosaurs,* "The World in the Age of Dinosaurs." (This second chapter has been written by Christopher A. Brochu and Colin McHenry):

"Until quite recent times, feathers were considered to be exclusive to birds, including *Archaeopteryx*. During the course of 1990s, however, several new discoveries from north-eastern China have proved otherwise. A fossil locality in Liaoning province began to yield the remains of small theropod dinosaurs, some of them with curious fibrous structures surrounding the body .... *Protarchaeopteryx* and *Caudipteryx*, the feathers are unambiguous – they have a central shaft (rachis) as well as fibers (barbs) .... We can no longer simply draw a line that separates bird and theropod; modern birds are clearly living members of the Dinosauria ..."

P.296

When I first read these statements, I was troubled by the words "proved otherwise." If the "strange markings" did indicate wings on dinosaurs and if RACHIS AND BARBS were unambiguous, then the theory that birds arose from dinosaurs presented a substantial case, *not a conclusive* case but at least a *substantial* one.

I marked in the above mentioned book – "these claims must to be researched." It pleases me that I found the answers to these problems (stirred up by the fossils of China) four years later [In all my purchased books, I have the habit of marking the present year on the fly leaf. The marked date is 2003. As I presently enter these thoughts into my word processor, the year is 2007. I am now ready to tell you what I discovered.

I was not aware of the fact that it was one year only, after the feathered-dinosaurs were found, the entire science of paleontology was *disconcerted or "ruffled."* We are indebted to Nicholas Comninellis, M.D., for his book, *Creative Defense*. One by one, the dinosaurs had their feathers plucked off. He writes:

"If people can have their 15 minutes of fame, so can dinosaurs. Most recently, the international spotlight has focused on a chicken –size fossil from northeast China, its body apparently fringed with downy impressions. For paleontologists who believe that birds evolved from dinosaurs, this specimen (*Mononychus*) seemed the ultimate feather in their cap …. [BUT]

"An international team of researchers that examined the Chinese fossil (*Mononychus*) now concludes that the *fibrous structures are not feathers.*"[Emphasis, mine]

P.167 [From the paper prepared by R. Monastersky, "Paleontologists Deplume Feathery Dinosaur," *Science News*, vol.151, May 3, 1997: p.271]

Also, a great deal of commotion was involved in the discovery of *Sinosauropteryx*. The *National Geographic* has a beautiful photograph of this fossil found with a mammal in its gut – Its last meal. Right above this exquisite fossil is the question:

"First Dinosaur with Protofeathers?

"*Sinosauropteryx prima* 'First Chinese dragon feather,' as its name translates, was found at Sihetun in 1996 and named for filaments to have covered its body. Most visible here as dark streaks rising off the hips toward the tail, these filaments may have been Protofeathers from which avian flight feathers evolved."

Ibid. Pp.78-80

On Pages 81-83 is a *restoration* of this same fossil but now the *Sinosauropteryx prima* is bedecked with colored feathers from dark brown to beautiful beige. Let me try to explain the difference between a reconstructed and a restored fossil. On the first series of pages, *Sinosauropteryx* is in the original matrix: thus, it is neither reconstructed nor restored. However, should the bones be removed from the surrounding substance; cleaned up by a preparator; put together by a construction biologist under the direction of a paleontologist; and put on display under supervision of a museum curator, the bones would then become a reconstruction (the bones have been put together).

Obviously, the best way to study the fossil is to take it the way it was discovered. This is the only way to make your scientific guesses based on what you *see* – the *empirical setting*, where nothing is altered.

So, how can one change anything by a reconstruction? Merely in this way: an *empirical* study can be changed into a *theoretical* study. For example, let's say a certain curator believes that a dinosaur such as *Sinosauropteryx* was warm-blooded. He will have this reconstruction arranged in a dynamic stance; the dinosaur will appear to be running with its arms extended forward and its tail raised up. However, let's say the curator is one who believes that *Sinosauropteryx* was cold-blooded. He would have the restoration demonstrate a more sluggish appearance; the tail would be down, the arms at its side, the legs close together, the head not on the alert, etc.

Both possible reconstructions would embrace the question of metabolism in small theropods. The empirical discovery has been misrepresented by its reconstruction and has been relegated, by its assumed posture, to become the object of a philosophical debate over the hot issue of whether small dinosaurs were more like lizards (Cold-blooded) or more like birds (Warm-blooded). Once a reconstruction has muscles, skin, and coloration, along with some other features, it becomes a restoration. A restoration is more likely to spring out of an artist's imagination than it is to arise from the observation of a scientist who has studied the fossil empirically.

The restoration of *Sinosauropteryx*, which is the object of our present attention, really goes overboard in the *National Geographic* depiction of a feathered dinosaur. The model is life-size: legs poised for running, arms extended, tail up, ready to pounce on a lizard and suggests that it was warm blooded.

Out of an article that covers twenty-six pages, *Sinosauropteryx prima* is mentioned on eighteen of them. I would say, at first guess, this fossil was very important to the theory of living up to its name – the "First Chinese dragon feather." But, once again, after the entire hullabaloo by Chinese paleontologists, the discovery was minimized in the following way:

"Another proposed reptile-bird 'missing link' was based on the fossil discovery of a creature called *Sinosauropteryx prima*. Initial investigators thought this animal had both feathers and reptile-like features. Their opinion was disproved only about one year later by several leading paleontologists who found that the alleged 'features' were simply fibers of collagen, the thing from which tendons are made." [The collagen fibers, which reinforce the tissue, account for one third of the weight of bones]

Nicholas Comninellis

Creative Defense

P.165

[Anonymous, *New Scientist*, 154(2077):13, April 12, 1997]

Jennifer Ackerman, the author of "Dinosaurs Take Wings" in the *National Geographic* which we are presently reviewing, reported *Sinosauropteryx* "one of the most important dinosaurs finds of the 20th century." She spoke of her memorable meeting with Ji Qiang (of China's National Geological Museum) and how she viewed this sensational find as a possible missing link between dinosaurs and birds. She even peered inside the silk-wrapped gift box at the stunning chicken size fossil with a halo of feature-like structures on its back and tail. Since all this was disproved a year later, apparently the feathers were purely a product of the imagination. It is funny how a phantom of evolutionary hope can materializes into the vivid reality of a *sensational find*; how illusory feathers of a dinosaur can be the harbinger of present day birds in all their authenticity. I wonder how Ackerman felt, after her report, a year later! I truly believe that such illusions of *Sinosauropteryx* and other species are quite possible due to the following reasons: Because of the growing recognition that birds are the descendants of dinosaurs and for the reason that such an evolution raises the possibility that dinosaurs had feathers, there is a strong desire for evolutionists to see at least some fuzz or protofeathers on such specimens.

I have been viewing the art work (restorations) of some of the fossils discovered in Liaoning, China. They are the product of artistic imagination and what is troubling to me; such drawings are passed on to the public as true science. Under the drawing of *Sinosauropteryx* is this statement:

"It is likely that several birdlike dinosaurs may have had feathers or similar structures, but, as they were not preserved, we know nothing about them."

Rocks, Fossils and Dinosaurs

P.392

Information coming out of the above mentioned book has been precise and clear-cut in describing *dinosaur feathers* but there is this inference from the above statement: some bird like dinosaurs had feathers since there *is evidence* and it is likely that other bird like dinosaurs may have had feathers even though there *is no evidence*. This one sentence is very subtle since it is implanted within our thinking, every time we think of a bird like dinosaur we should think of it as having feathers. And this in spite of the fact, feathers *were not preserved, we know nothing about them.*

The bottom line is this: *Sinosauropteryx* has been found by leading paleontologists not to have feathers but simply fibers of collagen. Have you noticed that these feathered dinosaurs continue to be defrocked right before our eyes? But we will consider a couple more – *Caudipteryx* and *Archaeoraptor*. Paleontologists know that *Archaeoraptor* is a fake so I will deal with that fossil later. I have a thing about fakes which may surprise you and make you wonder about my never wanting to take advantage of them so far as making the evolutionists look bad. But for now let's talk about *Caudipteryx*. This fossil should make every creationist sit up and take notice! – It actually had feathers and there is not a doubt about it. This fact is a demonstrable reality. "Well, that's it," you say. "All that you have written thus far is for naught. Why did you bother? It is true – Dinosaurs were the precursors of birds since they did have feathers." With forthcoming information, this response will quickly be quieted.

Caudipteryx ("feather-tailed") has been described as a bizarre theropod – an exciting event for paleontology because of its half wing. This creature has been originally interpreted as a feathered dinosaur *but it is a secondarily flightless bird.* Stephen Jay Gould has actually helped us out with the probable identification of *Caudipteryx*. He wrote a fascinating essay entitled "Tales of a Feathered Tail" in his book *I Have Landed* and pages 327-328:

" .... At least one genus – *Caudipteryx* ("feather-tailed," by etymology and actuality) – holds undoubted status as a feathered runner that could not fly. And so at least until the initiating tidbit for this essay appeared in the August 17, 2000, issue of *Nature*, one running dinosaur with utterly unambiguous feathers on its tail and forearms seemed to stand forth as an ensign of Huxley's intellectual triumph and the branching of birds within the evolutionary tree of ground-dwelling dinosaurs. But the new article makes a strong, if unproven, case for an inverted evolutionary sequence, with *Caudipteryx* interpreted as a descendant of flying birds, secondarily readapted to a running lifestyle on terra firma, and not as a dinosaur in a lineage of exclusively ground-dwelling forms (T.D. Jones, J.O. Farlow, J.A. Ruben, D.M. Henderson and W.J. Hillenius, 'Cursoriality in Bipedal Archosaurs,' *Nature* 406 [17 August 2000]).

"The case for secondary loss of flight rests upon a set of anatomical features that *Caudipteryx* shares with modern ground birds that evolved from flying ancestors – a common trend in several independent lineages, including ostriches, rheas, cassowaries, kiwis, moas, and others."

[Ibid. 328]

Although I have no reason for getting into a "case for an inverted evolutionary sequence", I will merely make mention of the fact that "inverted" means "to turn upside down" and in the case of *Caudipteryx*, it is surprising to see that this bird (from an evolutionary standpoint) was the first to renounce flight and so soon after birds took to the air (Gould). In other words, the term "inverted" is just another invented expression for evolutionists to explain the problem of how (in this case) a ratite showed up in the sequential order of evolutionary events. Flying birds were supposed to evolve from flightless bird and not the other way around. *Caudipteryx*, in cashing in its wings, was yet another reason why Gould gave up on progressive evolution and turned to the "fundamental nature of branching." In this way, Gould would not be surprised, in his understanding of evolution, that an "ostrich" could show up from time to time in the geologic column as existing contemporaneously with flying birds.

In any case, *Caudipteryx* is not a feathered dinosaur but a flightless bird and creationists would expect it to have feathers. Besides, this flightless bird was closer in proportions to a modern ostrich than to a bipedal dinosaur. Gould maps out three reasons why this is true – why the skeleton of *Caudipteryx* "falls into the domain of flightless birds rather than the space of a cursorial (running) dinosaur." In comparison with small dinosaurs this creature had:

*A relatively shorter tail.

*A center of gravity located in a more forward (headward) position.

*Relatively longer legs.

[Ibid. P.328]

[For those of you who seek further information on ratites please refer to my book "Was Darwin Wrong? Yes", Pages 42, 72-73]

Caudipteryx IS NOT A FEATHERED DINOSAUR but rather, A FLIGHTLESS BIRD.

I was very fortunate to have attended an international traveling exhibition, sixteen years ago, at the Natural History Museum of Los Angeles County. The exhibition was entitled "Dinosaurs Past and Present" and the Guest Curator was Sylvia J. Czerkas. I shall never forget the dinosaur paintings by Mark Hallett, Douglas Henderson, and John Sibbick. Also the Curator's husband, Stephen A. Czerkas who is an author, sculptor and paleontologist had some of his sculptural restorations of prehistoric animals on display. His restorations were simply beautiful but although he is a breath-taking sculptor, he made an abysmal purchase in 1999.

Jonathan Wells tells about it in his book, *Icons of Evolution*:

".... Stephen Czerkas and the National Geographic Society announced that a fossil purchased for $80,000 at an Arizona mineral show was 'the missing link between terrestrial dinosaurs and birds that could actually fly.' The fossil, which was apparently smuggled out of China, had the forelimbs of a primitive bird and the tail of a dinosaur. Czerkas named it *Archaeoraptor*."

P.124

Archaeoraptor was the main feature in the November 1999 issue of the *National Geographic*. Evolutionary scientists were thrilled to have a fossil demonstrating their theories and experimentation with flight.

At last! A realistic version of their "baby *Tyrannosaurus* with feathers" was found. But, once again, the scientific world would suffer a setback. The rush to the paleotological mines for the discovery of the fossil of gold, turned out to be the fossil made out of pyrite – "fool's gold." *Archaeoraptor* was a BIG FAKE.

Steven Jay Gould, in *I Have Landed*, describes the situation in this manner:

"Then, in June 1998, Ji Qiang and three North American and Chinese colleagues reported the discovery of two feathered dinosaurs from Late Jurassic or Early Cretaceous rocks of China ("Two Feathered Dinosaurs from Northeastern China," *Nature* 393, 25 June 1998).

"The subject has since exploded in both discovery and controversy, unfortunately intensified by the reality of potential profits previously beyond the contemplation of impoverished Chinese farmers – a touchy situation compounded by the lethal combination of artfully confected hoaxes and enthusiastically wealthy, but scientifically naïve, collectors. At least one fake (the so-called *Archaeoraptor*) has been exposed, to the embarrassment of *National Geographic* ...."

P.327

Wells brings out the fact that a clever forger had fabricated the fossil; the Chinese paleontologist Xu Xing had proved the specimen consisted of a dinosaur tail glued to the body of a primitive bird. [Pittack – "I do not fault the 'impoverished Chinese farmers' as much as I do the 'scientifically naïve.'"]

Wells also reports the shocking reaction by Storrs Olson, curator of birds at the Smithsonian Institution in Washington, D.C.:

"Olson blasted the Society for allying itself with 'a cadre of zealous scientists' who has become 'outspoken and highly biased proselytizers of the faith' those birds evolved from dinosaurs. 'Truth and careful scientific weighing of evidence have been among the first casualties in their program,' wrote Olson, 'which is fast becoming one of the grander scientific hoaxes of our age.'"

Icons of Evolution

P.125

How do I regard the entire situation? I think the forgery of the fossil was unfortunate for the cause of true science but especially for the cause of what I deem as the pseudo science of evolution. Not every evolutionary scientist should be indicted for this fake, fraudulent, and false fossil. As you just read in the above reaction of Olson, not every evolutionist should suffer the attack of creationists who want to take advantage of this situation.

Creationists made the following charge:

Archaeoraptor was *touted by scientists* as the dinosaur-bird transition (Sloan 1999). *This charge was answered in the following manner:*

First of all, the "Archaeoraptor was NOT A SCIENTIFIC FRAUD." [In fact, an evolutionary paleontologist was duped into making the purchase]

In the second place, was Gould certain that the Chinese worker knew the tail came from a separate fossil? [Pittack: I would not go as far as Gould went in making an evaluation that the fossil was an artfully confected hoax]

Thirdly, "*Archaeoraptor* was published in the popular press, not in peer-reviewed journals. The main author of the article about it was *National Geographic's* art editor, not a scientist."

In the fourth place, "*Nature* and *Science* both rejected papers describing it, citing suspicions that it was doctored and illegally smuggled (Dalton 2000; Simons 2000). Normal scientific procedures worked to uphold high standards."

The above four points were made by Mark Isaak in *The Counter-Creationism Handbook.*

P.134 [I think Isaak has answered the charge quite well. Most evolutionists are not careful in their criticisms of creationists but, at times, creationists are just as guilty in their analysis of certain situations regarding evolutionists]

Enough time has been spent on seeing how feathers have been placed on dinosaurs where they do not belong and understanding the reasons for their removal and being plucked out. Let's end up this section by getting a few scientists to agree with us. Some of the quotes, we are already familiar with their contents.

Alan Feduccia (University of North Carolina ornithologist), is "the best-known critic of the theory that dinosaurs gave rise to birds. He sees no proof that the dinosaur had feathers and doubts that any will be forthcoming. Feathered wings were 'the most complex appendage produced by vertebrates,' he says; 'it's implausible that an animal would have developed feathers if it did not fly'"

[Gibbons, Ann, "New Feathered Fossil Brings Dinosaurs and Birds Closer," *Science*, vol.274 (November 1, 1996), pp.720-721]

The Chinese fossils were the supreme hope of feathers on dinosaurs but the hope grew dim and then went out:

"An international team of researchers that examined the Chinese fossils now concludes that the fibrous structures are not feathers."

[Monastersky, R., "Paleontologists Deplume Feathery Dinosaur," *Science News*, vol.151 (May 3, 1997), p.271]

Philip Currie of the Tyrrell Museum of Paleontology, Alberta, Canada was at least unsure about the *pattern of feathers upon the "dresses" of the dinosaurs of China.* From what I understand, Currie can identify any species of dinosaur by the mere examination of a tooth. From the reading of his life and based on his work in the field, I know that Currie is a cautious scientist and not willing to advance conclusions that are not carefully thought out. He is unwilling to advance a foregone conclusion:

"They look so much like the feather impressions seen in the bird fossils at the same site that you can't come to any conclusion other than the fact that you're dealing with feathers." .... "Now, they may not be feathers. They may be featherlike scales, they may be hair, and they may be something else."

[Monastersky, R., "Hints of a Downy Dinosaur in China," *Science News*, vol.150 (October 26, 1996), p.260]

[They were identified as "something else" – fibers of collagen, the thing from which tendons are made – not feathers]

In *The Dinosaur Data Book*, David Lambert and the Diagram Group write:

"Feathered dinosaurs? *Below*: The small theropod *Syntarsus* is sometimes shown with feathered head crest and short feathers covering the body. Some scientists think all small theropods had an insulating covering of feathers. Were they warm-blooded, feathers would have helped them keep an even body temperature. Yet no feathered dinosaur is known apart from *Archaeopteryx*, usually thought of as a bird."

P.197

This paragraph was written eight years before the discoveries of the alleged feathered dinosaurs in China (1998). I doubt whether those discoveries would have altered the thinking of Lambert and especially since the information as to the true identity of the fossils was forthcoming so soon.

We already know about the angered statement of Storrs but we need to read it again:

"Truth and careful scientific weighing of evidence (that birds evolved from dinosaurs) have been among the first casualties in their program .... which is fast becoming one of the grander scientific hoaxes of our age."

Ibid. P.125

Since we have already removed feathers from dinosaurs where they do not belong, let us place feathers on *Archaeopteryx* and other birds where they do belong.

CHAPTER 10

THE *ARCHAEOPTERYX* ANSWERS THE QUESTION – AM I A BIRD?

(PART II)

(Do Feathers Make *Archaeopteryx* a Full-Pledged Bird in Its Own Right?)

"He who dwells in the shelter of the Most High will rest in the shadow of the Almighty …

He will cover you with his feathers, and under his wings you will find refuge."

Psalm 91:1, 4

NIV

Apart from figures of speech, such as the above verses, everyone knows that feathers and wings belong to birds but not everyone is certain that *Archaeopteryx* can be declared a bird even though bearing these same characteristics. If this is true and we know that it is, the only way that *Archaeopteryx* can stake a claim to its bird-ship and to be declared fully worthy of these biological attributes, is to be declared unreservedly that it is, without doubt, a true, genuine, and fully authenticated bird.

The most logical basis for determination of this fact is to begin with the first man to describe *Archaeopteryx* – "the fossil wonder of Solnhofen." Richard Owen was the superintendent of the natural history collections of the British Museum. Within a few months of the fossil's arrival in London, it was described in a scientific paper presented to the Royal Society. Evolutionists of Owen's day criticized him for paying more attention to the avian aspects of the creature and not enough time to the lizard-like features. But the judgment passed upon him was far worse in this day and time – some creationists, while Owens is resting in his grave, have the audacity to accuse this creation-scientist of plotting the so-called fake wings of *Archaeopteryx*. This has been already confirmed to be incorrect and we will, therefore, allow Owen to remain safely in his grave until that

Great Day when all men must give an account of deeds done on this earth (To God and *not men*).

As far as Owen putting the wrong emphasis on the reported fossil, that issue is entirely a matter of opinion. While describing this specimen as a bird *he did not omit the reptilian characters*.

David Norman in his book *DINOSAUR!* (Based on the acclaimed four-part television series hosted by Walter Cronkite) wrote about Owen's description. Norman does not quote Owen but he does give a general run down of his report turned in to the Royal Society:

"The specimen shows a beautiful long tail, down the centre of which runs a row of bones – typical of a reptile, for a bird has a short stumpy tail forming the "pope's nose." Fringing the tail on either side is a fan of clearly *preserved tail feathers*. The legs are long and slender, ending in *bird-like feet*. Near the front of the chest is a strong and well preserved furcula (wishbone), again *pointing very clearly to the creature being a bird*; and the *wings display a fine array of primary and secondary feathers as do those of any bird*. In contrast with birds, the hand has three well developed, sharply clawed fingers.... .... Owen concluded that this was *definitely an ancient bird*, and one that was extremely interesting because it showed a variety of *primitive vertebrate features which had become modified in later birds.*" [Emphases mine]

P.132

Readers! After noting the words I emphasized in the two above paragraphs, would you have a problem, if sitting in the leading chair of the Royal Society, in accepting Owen's report? With all the knowledge this man contained in his brain, how could he deduce anything else except the specimen that he had described was actually that of a primitive and extinct bird.

I like Steve Fiffer's comments on Owen's report – it's short, to the point, and is a fair assessment of the description of *Archaeopteryx* submitted to the high echelon of Britain's elite men of science:

"No doubt Owen, a superb anatomist, would have examined the specimen from head to toe – if part of the head hadn't been missing. As it was, he did conduct a thorough analysis of the fossil. Acknowledging that the *Archaeopteryx* did possess some reptilian features, Owen pointed to the feathers as 'unequivocally' proving that the creature was a bird. There was no evidence that this was one of Darwin's missing links, he reported to the Royal Society in November of 1862."

Tyrannosaurus SUE

P.48

I can visualize the Royal Society sitting at the round table for discussion of the *Archaeopteryx* and how its animal classification should be viewed. Listen, as these leading scientists have their discussion! [The following dialogue is, of course, pure imagination. I have no idea as to how the Royal Society conducted business back in the 1860s or, for that matter, any other time]

Chairman: "Do you think Owen is correct in his description?"

Secretary: "Do you mean if he is correct in calling this fossil a bird?"

Chairman: "Exactly."

Secretary: "I don't think so."

Chairman: "Why not?"

Secretary: "Did you ever see a long tail on a bird? – I know I have seen long tails on lizards and this creature is too much like a lizard."

1st Scientist: "I don't mean to contradict our eminent secretary in what you have said but I would like to remind you that we have also observed lizards with short tails. The point I'm making is this: Let's assume that the *Archaeopteryx* is a bird. If you are going to compare a long tailed bird with a long tailed lizard, then why not compare short tailed birds with short tailed lizards. Are all birds, in fact, lizards?

Secretary: "I'm not sure but I don't think we are talking about the same thing."

2nd Scientist: "I kind of side in with the secretary. I think that there is more to be concerned about than the mere tail. There is the matter of those claws – too much like a reptile."

3rd Scientist: "I think that I can clear up that problem. Last summer, if you remember, I was in South America. I didn't actually see a hoatzin but I have seen pictures of them and they definitely have claws. And something else – that bird collection that we have started in our museum of natural history, I call to your memory and attention, have a touraco and ostrich from Africa; both of these specimens have claws and both of them are birds – as you well know."

Secretary: "I don't intend to pursue this all day long but it is those teeth that bother me."

2nd Scientist: "Yes! What about that?"

Chairman: "I fail to see the significance of teeth in ancient birds. The teeth fail to demonstrate any connection of *Archaeopteryx* with any other animal since every subclass of vertebrates has some with teeth and some without. And I would like to finalize this discussion with these closing remarks: This bird and I do say bird, has the feathers and wings of modern birds. I do not know of its transitional history and, gentleman, neither do any of you. It simply has none and it is, in deed, a rare and unusual bird."

1st Scientist: "Didn't we read in the report of our prestigious and most esteemed Sir Richard Owen, the feathers are asymmetrical and their center shaft and parallel barbs are like those of modern birds?"

Chairman: "Yes, we did! And that gentleman ought to conclude our remarks for the day. All those in favor of accepting Owen's well written report and classifying this ancient creature as belonging to the Class Aves, keeping in mind that Classification must be have its day with the International Standards Committee, signify by your Ayes! Thank you gentleman and now let's drink a toast, before we adjourn, to the health of Professor Owen and to this honorable Society!"

There was perfect logic in Owen's identification. Certainly, *Archaeopteryx* was unique in his reptilian characters but the feathers and wings outweighed them and tipped the scales in favor of bird identification. No other animal had feathers at the time of Owen's identification and, in spite of theories from the 1860s involving Huxley and other evolutionists, coupled with the modern attempts to attach feathers to dinosaurs by Chinese paleontologists, still no other animals but birds can model the feathery clothing of nature.

The theory, dinosaurs were birds to be and birds were dinosaurs that were, is (as we learned from chapter nine)\, the most outlandish of modern hoaxes ever perpetrated on the gullible public. I have spent a great deal of time picking the plumage from off dinosaurs. Informed readers should appreciate the advice of the evolutionary professor,

John Ostrom of Yale University, who still believes in the connection of dinosaurs with birds but approaches the subject with sensibility:

"Now if are going to subsume birds into dinosaurs, you're going to alienate every ornithologist on earth. Why obscure them? Are you going to go out and dinosaur-watch this morning? .... Why is it necessary to confuse the question by saying [birds are] dinosaurs? A bird is a bird. And it came from a dinosaur-type ancestor. Specific ancestory is very much in question. I'm not claiming *Archaeopteryx* is on the main line of avian, but *I don't see any characteristic, any anatomical feature in Archaeopteryx that makes it impossible to have been an ancestor of later birds. I think the simplest explanation is that the two are related.*" [Italics mine]

Hunting Dinosaurs

Pp.114-115

Aside from the evolutionary distractions, this is probably a most sensible piece of information. I would like to pose a question for our consideration: Does it make sense to say that birds have feathers and therefore because *Archaeopteryx* is related to birds (the "simplest explanation") and also has feathers; *Archaeopteryx* is perhaps and just maybe a bird?

When Ostrom, an expert ornithologist, makes such a candid remark, "I don't see any characteristic, any anatomical feature in *Archaeopteryx* that makes it impossible to have been an ancestor of later birds," I hold to that perception that deems the "first bird" to be just that – a bird. Ostrom says, "A bird is a bird" and it is logical for me to add, "Of course, of course." An *Archaeopteryx*, because of its feathers and wings, is of course a bird in its own right.

The Royal Society did not receive a report from Owen regarding the skull of *Archaeopteryx*. The reason for that: there was no head found with the body. But later specimens proved that not only did it have a bird-like skull but a furcula (wishbone), as well. The "head was a complex structure; with paired nasal horn ridges and preorbital hornlets" (Gregory S. Paul). What is the truth of the matter? – *Archaeopteryx* is a bird. It had perching feet and wings of a bird, not to mention primary and secondary feathers to match the feathers of any modern bird. Everything points very clearly to the creature being a bird. And above all, it could fly.

It is ironic for this creature to be referred to as "old wing" when its wings were just as new as the very best that birds have to offer. At the present time, many scientists believe that it was an accomplished flyer. *Archaeopteryx* is tapping scientists on the shoulders to get their attention by saying, "I have an airfoil superbly adapted for flight. I am not a schizophrenic with three personalities – a reptile, dinosaur, and bird. I am only a bird with an "avian complex" – but not a "complex personality." A few scientists want to change the classification of *Archaeopteryx*. For example, Robert Bakker would prefer to "demote the Class Aves to a subdivision of the Class Dinosauria (or Class Archosauria with dinosaurs as a subclass)."

[See *The Dinosaur Heresies*, P.458]

Can you visualize birds not standing alone in their own division? [Picture *Archaeopteryx*: going to the airport where he has been hired as a pilot by the *Classification Airport System*! He goes to his locker to put on his flying jacket only to find out, as he looks in the mirror, something is different – the labeling over the left pocket has been changed. "And what's this!" He exclaims. The label now reads *Dinosauria* air lines instead of *Aves* air lines. Out loud he says, "Hey, I don't fly for this air line – it's a different 'branch' altogether." Just then, his boss comes out from the stall and says, "You do now. Class *Aves* no longer exists. And by the way, you have been demoted. You no longer stand alone and that will mean a decrease in your pay."]

As I said before, this subdivision will probably never happen. *Archaeopteryx* was once regarded as a feathered reptile or a lizard-tailed bird but is now classified as a bird. There must have been quite a few evolutionists that could not overcome the feather and wings to catalogue it in any other Class except Aves.

Edwin H. Colbert writes in *The Great Dinosaur Hunters and their Discoveries*:

"The discovery in Germany was of the first skeleton of *Archaeopteryx*, the oldest known fossil bird, and *a bird about which there can be no doubts,* because with the skeleton are the imprints of feathers in the fine-grained limestone." [Italics, mine]

P.40

Evolutionists do have their *doubts.* That is why they spend so much time on attempting to make other animals into birds, such as theropod dinosaurs. One gets the feeling that evolutionists are pressed into the bird classification of *Archaeopteryx* since they have a difficult time in explaining the many stages involved in the making of a wing. They have already failed miserably in explaining how feathers evolved from scales. The attempts to explain the piecemeal development of a wing by the slow development of chance mutations occurring within the cycle of natural selection, to me, is nonsensical. But I suppose with time on their side, anything becomes possible to the evolutionist. Adrian J. Desmond in *The Hot-Blooded Dinosaurs a Revolution in Paleantology* has this to say about wing development:

"St George Mivart saw that continued and progressive modification eventually giving rise to a fully integrated structure like a wing was *a near impossibility for Darwin's cherished Natural Selection.* Instead, Mivart was forced to consider that the wing and all its harmonious accoutrements sprang into being by a sudden jump, as if there were some internal directing mechanism."

P.134

When Desmond writes "a near impossibility" I am not positive he is quoting Mivart or writing for himself. In either case, it does not matter. All evolutionists are reluctant to say that natural selection could not possibly bring a wing into existence. Otherwise, they are left without the mechanism of change and variety which can supposedly fashion all things, including a wing. After all, in Darwin's bible on evolution, I believe there is a scripture that goes something like this: "In the beginning was Mutation and Mutation was with Natural Selection and Mutation was Natural Selection .... All things came into being through Mutation." Mivart was ready to allow for the miracle of "sudden jump" and "some internal directing mechanism" to replace the bible of Darwin. But aside from all this, he also would not permit God's direction in the creation of a wing inferred by John 1:1-3 (NIV) in God's Bible of Instruction:

"In the beginning was the Word, and the Word was with God, and the Word was God. He was with God in the beginning. Through him all things were made; without him nothing was made that has been made."

I can never bring myself to believe that men can replace one miracle with another miracle and come to grips with the deception that their miracle-of-evolutionary-replacement is better than God's original miracle of Creation. No small wonder that the evolution of the wing has remained an enigma.

God provided the answer for the mystery of wing development in the first chapter of Genesis, the Gospel of John, the book of Hebrews, and the epistle written to the church of Colosse. All these books contain the doctrine of creationism in their very first chapter. This is so mankind would easily remember where life first originated and be able to cite the references. But whenever men turn away from His explanation and seek for clarification from the false science of Darwinism or from their own false view; the "fully integrated structure" like a bird wing will forever remain an enigma and eternally, a mystery.

Desmond continues on with his writing and looks for an answer to wing development from the very thing that Mivart attempted to navigate around – PROGRESSIVE EVOLUTION. Desmond reasoned that "the process of taking to the air *must be a gradual one involving many stages* .... The answer to the riddle is that each of the components that in totality were to lead to *must originally have evolved for some purpose other than remaining aloft.* Only when the all the requisite structures were present could they be switched to a new function." [Italics, mine]

The Hot Blooded Dinosaurs

P.134

The reason Desmond uses the word MUST is because Evolution demands it. Otherwise, Natural Selection would simply crumble to pieces. To begin with, evolution calls for FAITH just like every OTHER RELIGION. It demands faith from its advocates. There are certain guidelines in evolution and one of them is natural selection. A person is left without a choice. Mutations, working within the sphere of Natural selection, are the mechanistic force behind variations and transformations in organisms – the only possible way for a wing to develop and come into existence. What other choice does one have except to accept this "irrefutable" dogma?

What is Desmond asking us to believe? Only that which stretches the credulity of man to the extreme limits, such as: countless random mutations, working through natural selection over a period of millions of years, will enable an organism to cope with its environment, to survive and outlive in neighbors in its struggle for survival, and all this being true each step of its transformation. Reader! Do you understand the complete and total futility of such reasoning? Each step must wait around for future steps in the final transformation of the given organism. Two steps must wait around for the third step, etc. etc. Each of the steps must then combine into a unit that is suitable to the needs of a particular organism to survive as the fittest.

Forgive me! But this is a perfect time to once again use our imagination. Picture the following conversation between two mutations:

Step One: "Good morning, Step Two! – How are you?"

Step Two: "Good morning and you must be Step One since you are calling me Step Two. How long have you been waiting for me?"

Step One: "Roughly, this side of two million years."

Step Two: "That's quite a long time. Aren't you fearful of rotting or coming to the end of your usefulness?"

Step One: "In response to the time element, I must say that a couple of million years are a drop in the bucket to evolution. And, goodness me! – I am not fearful of rotting. Haven't they taught you anything about mutations in the School of Variation? Mutations can be held aloft for millions of years in waiting for our brothers to form that perfect alliance, when we can serve our organism, by switching over to a new function."

Step Two: "I don't mean to be rude and perhaps I didn't learn as much as I should have during my stay in school but what function are we now serving? I mean: are we supposed to just wait here for other steps until the time we become a functional unit for our organism?"

Step One: "That's an excellent question coming from an average student. I see that you are in need of more patience. Our day will come and beside, we are doing some good right now."

Step Two: "And what is that 'some good'? We are merely a tenth of a feather or merely half a flying muscle. Will you please tell me how a fraction of a feather or a portion of a muscle can possibly be advantageous to our bird organism in his present struggle for survival and how that will help him to outlive his neighbor, production wise?"

I realize this isn't exactly the way evolutionists view the process of natural selection but they can not truly explain the piecemeal development of vital organs. It is difficult to get away from the nonsensical issue since only the finished product is functional. True, there is an attempt made to explain each mutation as being vital to the overall survival of the particular species involved. But it is a paradox to speak of *random* mutations and at the same time speak of each component evolving for a *specific* purpose. It is strictly amazing that Desmond can speak of mutations as not remaining aloft since they have evolved for some specific purpose, the scope of which is never quite explained and lacks sense and rationality. And how can they do otherwise but remain aloft? – They are not ready to serve until a group of mutations "decide" that they have enough for a forum, to be switched to their new function. With this in mind how can one escape the rationale of Mivart who was forced to consider variation as some internal directing mechanism? And how can mutations be saved up, over long periods of time, without losing their function for which they were originally and supposedly intended?

Desmond is oblivious to the fact that progressive evolution is a far greater miracle than even Mivart's "sudden jump" theory. Men will conjure up a *thousand miracles* to advance the theory of evolution but they will not humble themselves by accepting the *one miracle* of creation.

The odds against wing development happening over millions of years through random mutations for the *Archaeopteryx* is nil – the number 10 with 116 zeroes. But the wings came from some place and through someone. The Archaeopteryx fossils have no record of reptilian ancestors and no confirmation of avian descendents. However, aside from any mysteries which may surround the *Archaeopteryx*, the fact remains: this specimen had feathers and wings in its own right. The actuality of this bird's isolation and uniqueness in the so-called Triassic Period of the Geologic Time Table is an excellent example of God's special creation.

In the foot note, on page 192 in James F. Coppedge's book, *Evolution: Possible or Impossible* there is this interesting thought:

"The authors of the highly regarded volumes of *Avian Biology* wrote about *Archaeopteryx*: 'Unknown are the links connecting this momentous find to its reptilian ancestors on the one hand and to its avian descendants on the other.' Although tempted to draw conclusions, they said, 'Without paleotological support … conclusions must remain hypothetical.' (Ed., Donald S. Farner and James R. King, Vol. I [N. Y.: Academic Press, 1971], p.20.) It seems clear that there is no real evidence or proof that *Archaeopteryx* evolved from anything."

I have already drawn my conclusion – This bird is the result of genetics and by virtue of the fact that fossilized specimens are found without transitions, testifies for biblical creation of original gene pools in the beginning of time and this is regardless of any supposition of how the fossil record might have formed.

Let us continue in our pursuit of further evidence to confirm that *Archaeopteryx* had feathers and wings and is a bird in its own right.

As I said before, Gregory S. Paul is one of my favorite dinosaur artists and his descriptions in *Predatory Dinosaurs of the Word a Complete Illustrated Guide*, is top-notch. The following statement is a classic and I do not say this superficially and without thinking. If I did not know any better, seeing this account out of context and in its isolated form, would be easy to conclude that Paul was a creationist rather than an evolutionist.

This is a marvelous statement in confirmation of *Archaeopteryx* having feathers and wings and being a bird in its own right:

"In *Archaeopteryx* the wing feathers are fully aerodynamic, being the asymmetrical winglike airfoils found in modern flying birds. In a flight posture, *Archaeopteryx* looks reasonably streamlined and aerodynamic … … the wings surface/total weight ratio was well within the flying bird range, about equal to a crow's. The wing bones are also as strong as flying birds. The big furcula created by fusing the clavicles into one unit, helped lock together and immobilizes the shoulder. This is disadvantageous in most animals, which need as much arm reach as they can get. But it is a classical avian way of strengthening the shoulder for flight. And not just flight, but powered flight …"

Pp.217-218

This statement should remind us, in spite of the fact that *Archaeopteryx* has similar morphologies with the theropods, there is no such thing as a feathered theropod and the *Archaeopteryx* and other birds have nothing to do with the classification Theropoda. I have a right to disagree with those evolutionists who claim that feathers are a poor taxonomic character since there are other animals that *could have had them*. Science knows of no other animals that have feathers.

Archaeopteryx and other birds stand alone in the classification of Aves. The bird feathers and wings provide evidence of the fact; *Archaeopteryx* stands proudly alone and under the banner of Genesis 1:21-23, "So God created the creatures of the sea and every living and moving thing with which the water teems, according to their kinds, and *every winged bird according to its kind.* And God saw that it was good. God blessed them and said, 'Be fruitful and in crease in number and fill the water in the seas, and *let the birds increase on the earth.'* And there was evening, and there was morning – the fifth day." The *Archaeopteryx* has feathers and wings and is a bird in its own right.

We already know that the Solnhofen, *lithographic* bird was a great flyer but let us pursue this reality even further. We will presently enter our study on the flight capabilities of *Archaeopteryx.*

CHAPTER 11

FLIGHT CAPABILITIES OF *ARCHAEOPTERX*

(The "Ground Upward" Theory and the "Tree Downward" Theory)

GROUND UPWARD THEORY

RBP

11 a

"Like birds hovering overhead, the LORD Almighty will shield Jerusalem;"

Isaiah 31: 5

NIV

"And God said, 'Let the water teem with living creatures, and let birds fly above the earth across the expanse of the sky, So God created the great creatures of the sea and every living and moving thing with which the water teems, according to their kinds, and every winged bird according to its kind. And God saw that it was good.'"

Genesis 1: 20-21

NIV

Literature dealing with the flight capabilities of *Archaeopteryx* would fill quite a few library shelves. Therefore, in order to keep this format within a workable and sensible time block, a limited discussion will ensue concerning the two main theories of flight.

Towards the end of the 19th century, two theories about bird flight came on the scene. Samuel Wendell Williston believed in the "ground upwards" theory [This theory is named the "cursorial theory" and is often ridiculed and seemingly less probable than "the trees downward theory" – the "aboreal theory."] He accepted the idea that the ancestors of birds were like small theropods that were fast running. They would run along the ground, flapping their arms. The arms evolved into larger appendages and eventually into wings that would enable the creatures to become airborne.

Williston actually thought that the early form of wings were like propellers that would lift the "proto-bird" off the ground for, at least, a short hop. There are times when evolutionary thought borders on the fantastic. This was one of those times. Obviously, arms cannot spin like a propeller or produce a steady thrust.

This theory reminds me of life so long ago. Backtracking forty-six years, I used to take off in H-21 helicopters but never had the privilege of landing in one. As a sky diver, the act of sailing through space was an exhilarating experience. However, sitting on the bench before take-off, my mind would wander to such thoughts as: my French-cross must be executed perfectly, how close will I be able to guide my parachute to the X at the Drop Zone? Although the possibility of dying entered my thoughts, the fear of death was not one of my emotions. At take-off, as the blades whirled, there was a tremendous roar at lift-off, the dust swirled around, and the copter groaned as it prepared to set out into the blue skies of the early morning. Although having no fear, the thought crossed my mind – I must be a bit *crazy*. Imagination could not possibly have pictured for me the future when I would be writing a book on *Archaeopteryx* and considering the *crazy* thought of a scientist who believed that the lift-off for short bird hops could have been the result of arms of a living animal which spun like the propeller in a helicopter. But such are the things that make life interesting.

David Norman, a dinosaur specialist but a man, who is familiar with all areas of paleontological expertise, brings out two powerful arguments against the "ground upwards" theory. The following information is Norman's ideas condensed.

- Early "proto-wings" could not spin. They had to flap up and down. But this presents a physical problem for the theoretical sprinter. Arms stuck out from the body would cause aerodynamic lift and drag and slow the creature down. Drag would act like an air brake and impede the bird's movements.

- Modern birds also run to reach a suitable take-off speed because they already have fully functional wings with perfect aerodynamic properties for flight. Norman points out those early birds had neither perfect functional mechanics (bones, joints, and muscles) nor perfect aerodynamics.

Dinosaur!

P138

At this point, creationists would have to wonder how imperfect wings would be any kind of biological advantage in the evolutionary game of rudimentary-roulette. In other words, what part did the so-called "proto-wings" play in the "survival of the fittest" theory? How could mutations develop wings in the first place? Are mutations really accumulative? Could these genitival subtractions eventually add up to a perfect wing?

Most paleontologists dismiss the "ground-upwards" theory because of all the physical problems which make flight development a real quandary. Norman says, "How any creature can evolve the power of flight has long been a puzzle."

Ibid.P.138

The theory that hypothetically takes the place of the "ground-upward" supposition is no better.

Norman writes:

"This line of reasoning, which tends to suppose that bird ancestors developed feathers and wings because in some way they 'knew' that they could be ultimately used for flight, is also extremely suspect."

Ibid.P.138

Once the creation of birds is rejected, the whole scenario of flight becomes a major problem and especially since the various theories provide no solution but "stall-out."

John Ostrom, a well-known ornithologist, came up with a solution (?) to the cursorial theory (in 1974) by adding a few theories of his own. He gave a novel twist to the "ground-upwards" theory. He, like Williston, believed that flight must have originated among small, fast-running predators. He pictured *Archaeopteryx* which was much like a small theropod in its bone structure. Ostrom reasoned that it must have been fast running and, with its small, spiky teeth, was probably insectivorous. He visualized *Archaeopteryx* as about the size of a crow only with longer legs. Within the framework of this hypothesis, Ostrom developed his own suppositions about feathers and flight. Here are the deductions made by the evolutionary paleontologist and summed up by David Norman:

- Running at a high speed, *Archaeopteryx* would chase its prey and catch flying insects such as large dragon flies.

- This bird was, no doubt, endothermic and needed insulation and thus evolved a feathered edge to its scales.

- The final step for the Ostrom theory: The development of feathery scales that fringed the arms and would serve as an insect net. *Archaeopteryx* would leap into the air in pursuit of its kill before it flew out of reach. Eventually, the feather-fringed arms would produce a fluttering sort of flight.

Ibid. P.138

Michael Denton says that this theory is plausible to some degree but like other models it raises a number of problems. He points out " an obvious difficulty is that no known animal regularly catches flying insects by leaping after them ...Nearly all insectivorous vertebrate species take their prey on the ground ....One suspects that catching flying insects is more difficult than one might have imagined."

Evolution: A Theory in Crisis

P.207

Since most of evolution is based on "accidents," this theory had a strong appeal to evolutionists. Flight was not the ultimate intent of bird ancestors, but as an accidental development – a by product of the habit of catching and eating insects. At least this was the consensus of those scientists that took hold of this cursorial theory. Otherwise, one would have to conclude that animals were capable of predicting their own needs to meet the pressures of their environmental and ecological demands.

The creationist camp is often accused of upholding far-fetched theories when it comes to creationism and especially when they have an accompanying faith-factor to go along with their values and beliefs. But, in preceding pages of this book, we have been introduced to the absurd notion of how frayed scales were transformed into feathers.

Since the evolutionists have to come up with preposterous scenarios in order to save their theory, it should not be hard to understand why creationists find it difficult, if not impossible, to view evolutionists as being divorced from a similar "faith-factor."

John Ostrom abandoned the insect-trapping story in 1983. He said, "The insect net idea is dead. It did its job."

Luther D. Sunderland

Darwin's Enigma

P.74

But what job did the story do? – It deceived thousands of students around the world; birds evolved from reptiles and *Archaeopteryx* was, indeed, a transitional form. For nine years, this ridiculous theory was fully entrenched in the minds of students falling under the insidious nature of its influence. Perhaps my naïveté is too great not to trust the sincerity of the majority of scientists. This is why I am surprised at professionals who, in some cases, are willing to sacrifice truth and will stop at nothing when it comes to advancing their pet evolutionary theory. To openly admit that a certain theory "did its job" I can't help but think, is the epitome of a no caring attitude and a total disregard for all that is sacred in scholastic interactions with the public and with students looking to instructors for the truth of scientific convictions.

There is no credible, mechanical force in nature that would give support to the idea that a ground-running reptile evolved feathers from scales, accidentally frayed them, combined these features along with other traits which enabled a bird to fly. This is a far cry from science and more like the fairy tale that originates in a mind bent on manifesting antagonism towards creationism.

Since paleontologists – at one time – considered Ostrom to give a plausible account of the origin of feathers, perhaps it would be a good idea to cover this subject as a way of further review only with a little more detail.

Ostrom had conjectured wings first arose as a gadget to catch insects:

"It is possible that the initial (pre-*Archaeopteryx*) enlargement of feathers on its hand might have been to increase the hand surface area, thereby making it more effective in catching insects? Continued selection for large feather size could have converted the entire forelimb into a large, light-weight 'insect net.' It is not difficult to visualize how advantageous these paired 'insect nets' would be in snaring leaping insects or even in batting down escaping flying insects" (Ostrom, J.H.,1979, "Bird flight; How Did It Begin?" *American Scientists*, 67:45-56, p46).

We can all agree with Ostrom, "It is not difficult to visualize." No, it is not difficult to *visualize* almost anything proposed by almost anybody but what is difficult: almost nobody can explain the feasibility of almost nothing proposed by evolutionists as being advantageous. For instance, it is not difficult to *visualize* bird ancestors developing feathery fringes from scales for body insulation or scales becoming feathers as aids for flying. How simple it is to *visualize* the *advantages* of such suppositions! And how simple it is for all of us to *visualize* a hippopotamus ballet dancing but the mechanical aspect of such visualization is beyond practical explanation and forever beyond the possibility of belief.

Michael Denton helps us to understand why there must be much more than mere visualization:

"It is not easy to see how a impervious reptiles scale could be converted gradually into an impervious feather without passing through a frayed scale intermediate which would be weak, easily deformed and still quite permeable to air ……. ……. ….

"Take away the exquisite coadaptation of the components, take away the coadaptation of the hooks and barbules, take away the precisely parallel arrangement of the barbs on the shaft and all that is left is a soft pliable structure utterly unsuitable to form the basis of a stiff impervious aerofoil. The stiff impervious property of the feather which make it so beautiful an adaptation for flight, depends basically on such a highly involved and unique system of coadapted components that it seems impossible that any transitional feather-like structure could possess even to a slight degree the crucial properties."

Evolution: A Theory in Crisis

P.209

Duane T. Gish responds to the scales to feathers theory. He explains why feathers evolving from frayed-out scales should be considered pure fantasy:

"Scales are flat horny plates; features are very complex in structure, consisting of a central shaft from which radiate barbs and barbules.

Barbules are equipped with tiny hooks which lock onto the barbs and bind the feather surface into a flat, strong, flexible vane. Feathers and scales arise from different layers of the skin. Furthermore, the development of a feather is extremely complex, and fundamentally different from that of a scale. Feathers, as do hairs but unlike scales, develop from follicles. A hair, however, is a much simpler structure than a feather. The developing feather is protected by a horny sheath and forms around a bloody, conical, inductive dermal core. Not only is the developing feather sandwiched between the sheath and dermal core, it is complex in structure. Development of the cells that will become the mature feather involves complex processes. Cells migrate and split apart in highly specific patterns to form the complex arrangement of barbs and barbules."

The FOSSILS still say NO!

P.135-136

Gish is a creationist. Creationists are not supposed to be well informed about science or even sound like they are scientific. Since this is the case made by evolutionists, why is it that every time I compare the writings of a creationist with the writings of an evolutionist, the evolutionist appears to be writing for a child's book of fairy tales and the creationist comes across as dramatically more scientific in his descriptions and details of whatever scientific issue in which he is involved? The evolutionist simply leaves out a host of details in his writings even when writing for a scientific journal.

The reasons should be quite obvious. Whenever the issue concerns macro or major evolution, the creationist can afford to be more specific in writing in an empirical way because he does not have to bring in and surround his subject with philosophy and generalities. Whereas the "authentic" scientist, who attempts to explain evolution must (because of the nature of his belief) deal with generalities and use such terms as "maybe", "perhaps", and "quite possibly." The creationist writes from the standpoint of reality and of the things which he observes in nature, is more empirical in his descriptions. The evolutionist, of course, has never seen one kind of animal evolve into another kind of animal, has never seen a scale become a feather. He has no other choice but to write from the standpoint of philosophy. His writings become the only "evolution" evident – the *transformation* of reality into pure invention; the *change* from empirical knowledge into unobserved data which exists mainly in the conjectures and wild assumptions of the evolutionist; the *modification* of plain facts into the modicum of truth coupled with the hodgepodge of error. Even when the creationist talks about beliefs of creation, there maybe philosophy mixed in with truth but the developed facts fit the model of creationism much better than when the evolutionist takes his gathered facts and attempts to fit them into the model of evolution.

[Pittack's Note: I must make some clarification concerning the writings of evolutionists. I have read a wealth of material from their writings including all the essays of Stephen Jay Gould. I have read many books on paleontology, biology, and geology. I do not mean to imply nor is it my intention to disparage the entirety of their writings. Who am I to make such judgments? And besides, creationists share the same love of science only from different perspectives. It is only when evolutionists are attempting to describe the assumed macro-evolutionary processes of the biological world, my criticisms apply. For whatever technical abilities and writing skills are demonstrated in their writings on this issue: the end result will always be a fairy tale steeped in philosophy and exuding an enchantment far removed from the real world of science]

In spite of the salient facts shown in the above statement of Gish, evolutionary paleontologists still hold to the belief that feathers developed from scales. If you say something over and over; it gradually becomes a fact. For example, a scale evolving into a feather is a theory; a scale evolving into a feather is a possibility; a scale evolving into a feather is a mathematical possibility; a scale evolving into a feather is a mathematical probability; a scale evolving into a feather is a fact. Evolutionists, who believe in the blind forces of chance and genetic mistakes accounting for the change of a simple horny plate into the complex structure of a feather, must have an active faith coupled with a strong denial of truth. I will never be able to decide whether their faith is stronger in evolution or in their résistance to the facts of reality.

Stephen Jay Gould writes about feathers as modified reptilian scales:

"Let us return, as we must, to the classic case of wings. *Archaeopteryx*, the first bird, is as pretty an intermediate as paleontology could ever hope to find – a complex mélange of reptilian and avian features. Scientists are still debating whether or not it could fly …. ….

"Proto-wings have been reconstructed as stabilizers, sexual attractors, or insect catchers. But the most popular hypothesis identifies thermoregulation as the original function of incipient stages that later evolved into feathered wings. Feathers are modified reptilian scales, and they work very well as insulating devices."

Bully for Brontosaurus

Pp.144-145

Look at the word, "incipient"! Gould claims that proto-wings went through "incipient" stages and later evolved into feathered wings. In other words, such a change was merely "basic," "fundamental," or "elementary." Again, an evolutionist takes as impossible idea and simplifies it so that it becomes more acceptable to the unwary. The process of turning scales of a reptile into the feathers of a bird is no problem. In fact, it's elementary to at least this evolutionist.

Scientists, in their imagination, form explanations for things in the natural world. However, their rationalization is not always true science based on experimental facts. Rather, it is simply philosophy and speculative thinking. The apparent fact is this: the feathers of birds did not evolve by chance or random happenings in geological time periods. The four types of bird feathers are complex and wonderfully designed features.

The second theory of bird flight will now come to our attention – the "tree downward" theory.

In 1880, Othniel Marsh proposed another theory about how bird flight originated. He believed that the ancestors of birds could have possibly been tree-dwelling animals. The following are his suppositions and briefly described by David Norman and further reduced by me:

*Creatures that were tree-dwelling and lived in such a precarious habitat; evolution favored any characteristic that would tend to break their fall should they loose their footing.

*Ancestral birds might have evolved through those animals which went through various steps in developing parachute-like structures to break their fall.

*Feathery scales could have evolved gradually and helped break the fall of small, tree-climbing creatures. [Pittack's note: I do not know why "mother" nature did not teach her animals how to make good PLFs (Parachute Landing Falls). All an animal would have had to do is observe the direction of its fall and land on its muscle calf, hip, and its shoulder and flip over. With this perfect *tree-point landing*, the animal would have avoided injury from *tree-falls* by learning *landing-falls*]

*Once the bird ancestors developed a sort of limited parachuting ability by their feathery scales, then they gradually perfected the ability to glide from tree to tree. [Pittack's Note: Meanwhile during the first millions of years, we can be thankful that the production-and-survival-rate was greater than the accident-and-death-rate of these animals falling out of trees before they went "through various steps in developing parachute-like structures to break their fall."]

*From here, Marsh "glided" home with the final phase of his theory. In his imagination, he pictured ancestral birds leaving the gliding stage and finally going into active flapping flight. [Pittack's Note: I wonder how many millions of years it took before a gliding reptile suddenly became aware of the fact that it was a bird going into active flight by flapping its wings!]

Dinosaurs!

David Norman

Pp.139-140

This theory is filled with many "perhaps' and "maybes." Again, the stages of flight are based on guesses only. We all realize that a good hypothesis is a natural precursor of scientific laws and a possible harbinger of truth. However, there is not the slightest evidence or experimental facts to support these various points concerning bird flight taking place outside Othneil Marsh's fertile imagination. This is yet another example of a theory void of supporting facts; still Marsh's idea was widely accepted by evolutionists. This "Tree Downward" hypothesis is easier to accept then the "Ground Upward." – Why?

*Take off is a great deal easier when an animal launches itself from a tree.

*Gliding is more logical in its development than ground take off. Gravity creates speed for a gliding animal.

*Also, the body contours create some lift during descent so the gliding becomes possible.

In writing of the "up" and "down" theories, David Norman has this to say:

"The origin of feathers seems relatively uncontroversial: both 'up' and 'down' theories can accommodate it. It seems perfectly possible, if unprovable, that feathers developed initially as scales with fringes on."

Ibid. P.142

Amazing! – The feather development concept is almost at the place where it can hardly be contested. After all, both the "up" and "down" theories can accommodate it. What a strange piece of evolutionary reasoning! Norman states that although feather development is "unprovable" is still "perfectly possible." [Where does one draw the line between what is *perfectly possible and perfectly impossible*? Especially, when the only criterion is *improvability*?] I ask to be spared from this type of reasoning. And not this only but from the one theory that is used to support two other theories which I will now mention. For me; evolution from top to bottom remains a massive hodgepodge of unproven verbal emissions. In the above statement by David Norman, he uses one theory (the origin of feathers) to support two other theories (the "up' and "down" theories regarding the origin of flight). These three theories are used to support yet a fourth theory called evolution which is accepted by the scientific community as an authentic and indisputable fact. Does it get any worse than this! I am sometimes dumfounded by jargon that is nothing more than scientific chicanery. I should trust that my readers are very much aware: the transition from gliding to active powered flight cannot be explained by evolutionary theory.

David Norman is candid enough to admit there are many difficulties connected with the theory of flight. He either does an "about face" in this admission or he simply adds further reasons in stretching our credulity for understanding how the equation of feather evolution, coupled with flight evolution, plus the "difficulties" he will now mention, can possibly equal the *perfectly possible*:

"The difficulties created by active flying include the requirement for powerful muscles to power the wingbeat, a strong but very light skeleton, a powerful heart and lung system to supply oxygen and food to the flight muscles; and a highly sophisticated sensory system and brain to control and adjust the flight path of the animal at all times. This is a daunting list of requirements, and must all provide simultaneously if the 'flying machine' is to work."

Ibid. P.141

[Pittack's Note: Norman has just described one of God's creations. It's called A BIRD]

Yet, when the smoke settles, the gliding theory still takes front and center in the mind of most evolutionists. But when facts are non-existent, at what point does faith come into view? How are scientists able to explain their willingness to accept ideas which have no scientific basis other than to admit they resort to the faith modicum? Surely, faith is a part of the evolutionists' belief system.

Let us talk about the flight capabilities of *Arxchaeopteryx*. The big question is this – Could *Archaeopteryx* fly? Since *Archaeopteryx* is considered a missing link, evolutionists play down its flying abilities. The reason for this stance is quite obvious: A transitional form between reptiles and birds would not be expected to be a good flyer. In the evolutionary scheme of things, gradual steps eventually lead up to modern bird flight skills. We have observed that flight development is purely guess work on the part of the evolutionist.

Birds are among the lightest and the most fragile of vertebrates. They are rarely preserved as fossils and therefore futile for evolutionists to claim they are able to decipher the history of birds. The great mystery that surrounds *Archaeopteryx* is the fact that it was already a true bird with primary and secondary feathers, capable of flight. But these are not the only characteristics. Here are the known facts:

FACT ONE – It can no longer be argued that the *Archaeopteryx* lacked a bony sternum and was, therefore, a poor flyer. Gish makes the following statement:

"The fossils of *Archaeopteryx* were found in the Solnhofen Plattenkalk of Franconia, Bavaria ....The seventh specimen from Solnhofen was reported in April, 1993. This seventh specimen is remarkable in that it includes a bony sternum."

The Fossils Still Say No!

Duane T. Gish

P.132

Since the discovery of the seventh *Archaeopteryx* revealed a sternum not seen in other specimens because of the delicate nature of their anatomy; there was a place for the attachment of flight muscles and no reason for assuming *Archaeopteryx* not to be an excellent flyer.

FACT TWO – The flight feathers of *Archaeopteryx* are the most beautiful of biological adaptations.

Michael Denton writes:

"Flight feathers are remarkably light and strong and anyone who has played with one will know how easily a ruffled feather can be repaired merely by drawing it between the fingers. In addition to its lightness and strength the feather has also permitted the exploitation of a number of sophisticated aerodynamic principles in the design of the bird's wing."

Evolution: A Theory in Crisis

P.202

Denton mentions two other facts.

FACT THREE – "One problem common to all airfoils is turbulence, which reduces lift and causes stalling. Turbulence can be greatly cut down by the provision of slots in the aerofoil which let through part of the air stream and tend to smooth down the flow. Aeroengineers have used this principle by placing a small subsidiary aerofoil in front of the main wing ...."

Ibid. P.202

FACT FOUR – "The use of feathers also provides the bird with an aerofoil of variable geometry so that it has the ability to vary the shape and aerodynamic properties of its wing at take-off, landing, and for various different sorts of flight – flapping, gliding, and soaring."

Ibid. P.203

God is the Master Aeroengineer and not natural selection when it comes to manufacturing feathers! Certainly *Archaeopteryx* could easily have soared through the air with the greatest of ease. It had all the right combinations for flight: a bony sternum, fully aerodynamic wing feathers, asymmetrical winglike airfoils, strong wing bones, and the wing's surface/total weight ratio was correct for a flying bird.

The book of beginnings informs us that God created "every winged bird according to its kind" as He said, "let the birds increase on the earth" (Genesis 1:21-22 NIV). The Creator could not be any clearer in this passage of scripture: birds "were created together by the Lord's command; not one species evolved from another" (*The Battle for the Beginning* by John MacArthur, P.129).

I became a bird watcher at the age of seventeen. Every morning, before college classes began, I would go down to the woods with my note book and binoculars. I was fascinated by the colors and the different varieties of birds living in the habitat of Takoma Park, Maryland. As I look back on those days, I can't help wondering about the fact that many ornithologists, after years of observation, can still remain evolutionists. I would think that students of the bird culture would be the first to be aware of the wisdom of God and His love for beauty and design.

Creationists believe birds were already geared for flying. Evolutionists call the bird's anatomy, its ability to fly, and its remarkable migrational trait, evolution by chance convergence. Creationists call it the miracle of design and creation. *Arxchaeopteryx* did not suffer the mental anguish of being three personalities – a reptile, dinosaur, and bird. It had only an avian persona. *Archaeopteryx* was actually a created and then became an extinct bird. Thus reminding us of God's love in creation and man's fall which brought on the Great Flood. The Flood is our topic for discussion in the next chapter which contrasts the fast process of Catastrophism with the *alleged* slow course of action in the doctrine of Uniformitarianism.

CHAPTER 12

THE PLACE OF *ARCHAEOPTERYX* IN THE GEOLOGIC TIME SCALE

(Is *Archaeopteryx* in the Proper Time Zone and Apposite Order between Reptiles and birds?)

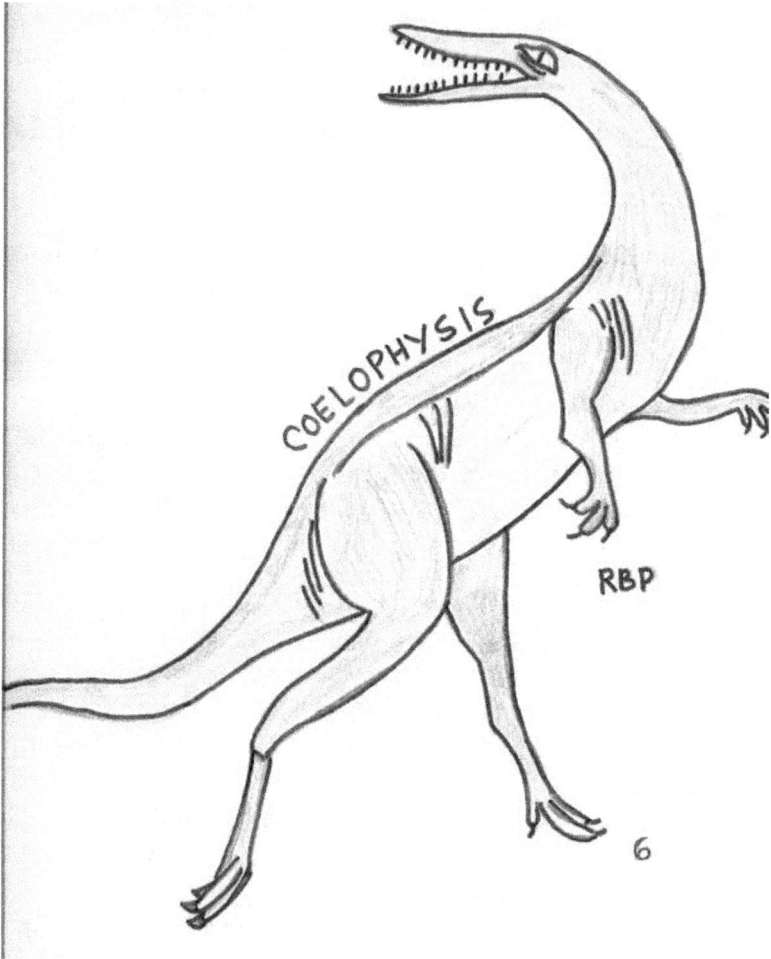

"To me this is like the days of Noah, when I swore that the waters of Noah would never again cover the earth. So now I have sworn not to be angry with you, never to rebuke you again."

Isaiah 54:9

NIV

"As it was in the days of Noah, so it will be at the coming of the Son of Man. For in the days before the flood, people were eating and drinking, marring and giving in marriage, up to the day when Noah entered the ark; and they knew nothing about what would happen until the flood came and took them all away. That is how it will be at the coming of the Son of Man."

Matthew 24: 37-39

NIV

	Events In Bakker's Time Line
5	PALEOCENE (next epoch) Totally modern, toothless birds. Birds without teeth - bills
4	CRETACEOUS (next higher strata) More modern birds with teeth. Marsh's toothed birds More advanced then Archaeopteryx.
3	JURASSIC (next higher strata) Primitive birds with teeth – Archaeopteryx.
2	TRIASSIC (next higher strata) Advanced "Reptiles" (bird-like Dinosaurs) with bird like bodies.
1	PERMIAN (next higher strata) The first primitive reptiles. Rocks from the Coal Age and Permian Period.

Were it not for the evolutionist Robert Bakker, this chapter never would have been written. In 1989 Bakker indirectly led me into the study of Dinosaurs. He alone was responsible for nurturing my fascination for these fabulous creatures; fashioned on the sixth day of creation – an event which I recognize but Bakker, of course, does not.

While Bakker quickened my love for dinosaurs when I read his book, *The Dinosaur Heresies,* he lacked the power of persuasion. If anything, he made me more committed to the cause of creationism and more determined than ever to support its philosophies and ideals. Because Bakker views birds as evolving throughout the Geologic Time Scale, he writes:

"Even in our own day, the Creation Science group in California grinds out pamphlets bearing the same message: One warbler 'species' might transform into another, but all birds have always been true birds and the first sprang full-blown from the creative hand of God."

The Dinosaur Heresies

P.301

Bakker must be a superficial reader if he thinks that this is the only message in creationist pamphlets. Perhaps he has glanced at a few pamphlets but his having read a book such as Gish's "Creation Scientists Answer Their Critics" is not very likely. Bakker has a shallow understanding of the doctrine of variation with limitation – an empirical observation as opposed to his teaching of variation without limitation which is highly philosophical and far removed from the realm of experimental observation either in the laboratory or in the field of nature.

He has always been critical of Genesis 1:1 but this one passage is more scientific than any portion from the writings of Darwin could ever hope to be.

John MacArthur reminds us:

"In the beginning" – that's *time.* "God" – that's *force.* "Created" – that's *action.* "The heavens" – that's *space.* "And the earth – that's *matter.* In the first verse of the Bible, God laid out plainly what no scientist or philosopher cataloged until the nineteenth century ...

"It's HARD TO IMAGINE anything more absurd that the naturalist's formula for the origin of the universe: *Nobody times nothing equals everything.*" [Emphasis, the author's]

The Battle for the Beginning

Pp.41 and 31

Returning to Bakker and the Geologic Time Zone, he makes this audacious statement:

"The stratigraphic proof for a Darwinian origin of birds appeared incontrovertible – the rocks preserved the stages of development in the exactly proper sequence through time. Any impartial observer might conclude that if God had really created birds, he must have been going out of his way to fool humanity into believing in evolution."

Ibid.P.303

There is plenty of evidence why God has left no doubt about the insidious nature of Bakker's remarks. We shall give confirmation why such remarks only place man in the position of self-deception and in the face of God's general revelation in nature and special revelation in the Bible.

At the very beginning of this chapter, I have made a chart so that my readers will be enabled to follow Bakker's line of argument for the correct placing of birds (including *Archaeopteryx*) in the Geologic Time Scale. Bakker asserts in order for a paleontologist "to accept an animal as a real 'missing link' between classes, the fossil is not only to display an anatomical structure intermediate between two distinct classes, but it also has to appear in the 'correct' sequence of time, intermediate between the two classes."

Ibid.P.303

To back up his claim with alleged *proof*, he demonstrates how the origin of birds lines up in correct sequence throughout the geologic column beginning with the lowest level and working up to the highest level. In the evolutionary scheme of progressive transmutation, one must begin with the deepest level since it is logical to assume that species recovered from farther down in the column will be the oldest in geological time. Look at the chart that has been prepared for you and you will see that number (1) begins at the bottom of Bakker's time zone! The following is Bakker's *correct sequence* for the origin of birds:

(1) The first primitive reptiles then (2) advanced "reptiles" (dinosaurs), then (3) primitive birds with teeth, then (4) more modern birds with teeth, and finally (5) totally modern, toothless birds.

Ibid.P.303

	Geologic Column Time Zones	
ERAS	PERIODS	EVENTS

CENOZOIC (recent life)
 	
65 million years ago until present
	QUATERNARY:
	  Recent Epoch
	  Pleistocene Epoch
	TERTIARY:
	  Pliocene Epoch
	  Miocene Epoch
	  Oligocene Epoch
	  Eocene Epoch
	  Paleocene Epoch
		(5)Truly modern birds

MESOZOIC (intermediate life) 225 million years ago to 65 million years ago	CRETACEOUS JURASSIC TRIASSIC	(4) More modern birds with teeth Archaeopteryx – (3) Primitive birds with teeth (2) Advanced reptiles (Dinosaurs)
PALEOZOIC (early life) 570 million years ago to 225 million years ago	PERMIAN PENNSYLVANIAN MISSISSIPPIAN DEVONIAN SILURIAN ORDOVICIAN CAMBRIAN	(1) The first primitive reptile

Bakker, to make certain the proper sequence is being followed, gives further explanation:

"Now, primitive reptiles had been found low in the strata, in rocks from the Coal Age and the Permian Period. Dinosaurs with birdlike bodies made their entrance in the next-higher strata of the Triassic Period. *Archaeopteryx*, a very primitive bird with teeth, showed up in the next strata, the Jurassic. Marsh's toothed birds had been more advanced than *Archaeopteryx* and appeared in the next period, the Cretaceous. And, finally, truly modern birds without teeth made their debut at the very end of the Cretaceous and the beginning of the next epoch, the Paleocene. *So it all fell into place exactly as evolutionists would have predicted.*" [Emphasis, mine]

Ibid.P.303

Everything seems so logical and precise to Bakker. Is the fossil record within the geologic column that consistent and so easily defined? Does his argument of *sequence* have any validity? Is there any truth to *geologic ages*? Does the circumstantial evidence of fossils give certain attestation and "prove" evolution to be a scientific fact? This chapter will attempt to make clear why every one of these questions should be answered in the negative.

It is necessary, first of all, to set up the foundation and framework of the evolutionary structure before dismantling it or tearing it down. Since we are going to be discussing geology in the main, it will be beneficial to focus in on what branch of geology will have our concern and attention. In its simplest form, geology is the study of the earth. In its complex form, it evolves six branches of study: (1) Physical geology (2) Historical geology (3) Economic geology (4) Structural geology (5) Geophysics and finally (6) Physical oceanography. We will be looking at physical, historical, and structural geology but not, of course, in great detail.

When I made numerous field trips throughout the region of California, I made certain to bring the "Dictionary of Geological Terms" (Third Edition) prepared by the American Geological Institute and edited by Robert L. Bates and Julia A. Jackson. When a creationist, like me, studies geology on his own, it is necessary to rely on some of the *tools prepared by evolutionists*. As far as I know, no dictionaries have been written up by geologists of the creationism philosophy. I also took along road-side field guides that described geology in different parts of

California. [I studied geology for many years but only as an arm-chair reader. I finally decided that the only way to really study this subject was to combine the photographs in books (along with their footnotes and commentary)with first-hand examination of the geology of various landscapes and by observing road-side strata exposed at those sections of highway that had been developed by blasting through mountains and hills] I would stop at road-side turn-offs, study my field guides; take pictures that I would eventually have made up into slides, made constant use of the dictionary in identifying the geological formations described in field guides; when it was legal, investigate these formations close up; when my slides came back from the Developing Center, made them up into programs. You would be surprised at how much geology you can learn in only about four or five years but still, this is barely enough time to scratch the surface of such a complex and multifaceted subject. I also went to the book store of Antelope Valley College and purchased the introductory study outline for the course in geology. I suppose it would have been easier to take a college course in geology. However, it wouldn't have been nearly as much fun and excitement as it is when I was able to study geology on my own time and scheduling; especially when I brought along one or more of my five sons to accompany me. I also made certain to take along an ice chest with cold drinks and my famous, delectable lunches.

I have read many books on geology written by evolutionists. Also, I read some books and articles on geology written by creationists such as George McCready Price, S. A. Austin, Harold G. Coffin, John Woodmorappe, and Harold W. Clark. Being a creationist I found it difficult, in the early part of my endeavor to learn as much geology as was possible; to separate the purely physical and structural geology from the historical geology. The reason I had for finding it difficult to study geology? The evolutionary authors of geology books always mixed descriptive and structural geology with the philosophy of uniformitarianism in their attempt to describe historical geology. I became more adept at straitening out this problem as the years went by.

In the margins of books from my personal library, I wrote out questions for evolutionists and creationists along with cross-references. I was always grateful to learn from these men and appreciated their information. I say this in full knowledge that a few eyebrows will be raised by my fellow creationists in response to my remarks. However, I am sure of their relaxing when I say that the philosophical foundation of religion, creation, and the Great Flood was dear to my heart. At the same time, the only thing that I could possibly take away from the foundation of evolutionists was the *descriptive nature of geology*. I say this because the creationist geology was based on a model which I found to be in harmony with empirical science as opposed to the evolutionist model based on uniformity which was not in agreement with empirical science.

Since uniformity or uniformitarianism is the assumed basis for evolutionism, it needs to be defined and made clear why it is a drawback for understanding the nature of events in the Geologic Time Column. I will give the definition taken from the above mentioned book, "Dictionary of Geological Terms" and page 546:

"The fundamental principle that geological processes and natural laws now in operating to modify the earth's crust have acted in much the same manner and with essentially the same intensity throughout geologic time, and that past geologic events can be explained by forces observable today; the classical concept that 'the present is the key to the past'. The doctrine does not imply that all change is at a uniform rate, and does not exclude minor local catastrophes."

This is the usual definition of uniformitarianism found in most evolutionary literature. On the other hand, the very best elucidation of uniformitarianism comes from the viewpoint of creationists, is the following taken from a book on creationism:

"The geological dogma of uniformitarianism is usually associated with the name of Darwin's fellow scientist and mentor, Sir Charles Lyell, although – like evolutionism – uniformitarianism is really a very ancient belief of pagan philosophers. The famous slogan of this doctrine is that 'the present is the key to the past.' The assumption is that present processes (erosion, sedimentation, volcanism, glaciation, etc), operating as they do at present, are sufficient to explain all the geological and other features of the earth.

"This uniformitarian assumption, if valid, would mean that the earth must be very old! Since no instance of true 'vertical' evolution has ever been observed during all of documented human history, the evolutionary time span needs to be almost infinitely great in order to make 'particle-to-people' seem even remotely feasible."

SCIENCE & CREATION (volume two)

Henry M. Morris,

John D. Morris

P.20

Uniformitarianism is usually associated with Darwin and Lyell since its focus is on the equation of evolution. The formula would appear this way:

NATURAL SELECTION + TIME = EVOLUTION

That is, DARWIN + LYELL = EVOLUTION. Darwin supplied Natural selection (Survival of the Fittest) which is believed to be the mechanism or driving force behind evolution and Lyell supplied the long periods of time or "deep time" that would be necessary to make evolution at least a feasible and workable philosophy.

It is interesting to note what the first of the above definitions of uniformitarianism has to say in the final part. It is almost like the portion is an after-thought placed in the original definition. The segment states: "The doctrine does not imply that all change is at a uniform rate, and does not exclude minor local catastrophes."

Apparently, it was through field observation that geologists came to the conclusion that geologic events contained evidences of a cataclysmic nature. In other words, they were forced to come to the conclusion that they had to give in to what their eyes were telling them; there were such things as local catastrophes. This is a no-brainer but the definition does not go far enough. George McCready Price wrote a book entitled *Common-Sense Geology*. Common sense should dictate that not only were there local catastrophes but the Geological Column was, itself, ONE GREAT CATASROPHE. This seems to place certain observers of geological events, in total denial of the true facts. Certain other observers turned away from uniformitarianism and accepted such things as regional and global catastrophes.

Allow me to explain, very briefly, why this predicament has arisen in the camp of evolutionary geologists. At one time the pendulum swung to the side of creationists and it was mostly accepted that the world was once destroyed by a Global Flood recorded in the Bible (Genesis 6:1 - 9:17) and called by creationists, the Noachian Flood. But along came Darwin with his *Origin of Species* in 1859 and greatly influenced by the writings of Lyell recorded in the *Principles of Geology*; the pendulum swung the other way in favor of the evolutionists. I urge you to remember that Darwin was not only a naturalist but a geologist as well. He understood the work of Lyell and was deeply indebted to Lyell for furnishing him with the alleged DEEP TIME – millions and millions of years to carry out the slow and methodical processes of transmutation of animal species and plant varieties. Evolution of any kind naturally implies continuity and there was no discrepancy between evolution and uniformitarianism. Evolution claimed that species were transformed from an original ancestor and descendants became different kinds of animals as time went by. Uniformitarianism simply supplied the essential time that was necessary for the progression of evolution. After all, uniformitarianism was itself, a form of evolution.

Uniformity and evolution went hand in hand in the nineteenth century and many geologists were turned away from the Bible and the Flood concept. Scientists listened while the "earth spoke" to them but it was not the earth that had spoken but Lyell. The Bible stated that the earth was deluged by the forces of water. Lyell spent an entire life time denying the book of Genesis and taught that the earth was not destroyed by water but taught that water was only a source of erosion and other geological processes that required long periods of time. Thus many people were aligned on two contending sides – Uniformitarianists and Catastrophists. Over the years, these two camps have turned into different names and different philosophies. Evolutionists have claimed some of the creationist positions and I am sad to say that many creationists have claimed some of the evolutionary teachings.

[Pittack's Special Note: I am not concerned with the divisions of Uniformitarianism such as its various branches. There is *substantive uniformitarianism* and *methodological uniformitarianism*; *new catastrophism* or *neocatastrophism* which includes "episodic sedimentation." I believe a work-up of these various items would only confuse my readers who, like me, are not professional geologists] The bottom line is this: Many scientists are including catastrophes in their stand on evolutionary geology. This stance is called *new catastrophism*. Some of these scientists lean so far over the edge in offering examples of catastrophism; they are often confused with creationists who believe in the Bible, especially Genesis, and in the Noachian flood. Creationists often find ammunition to support their belief in the biblical World-Wide Flood by quoting from the sources of men like Derek V. Ager who wrote the book, *The New Catastrophism*. Creationists like to quote Ager since his writings are bent against uniformitarianism which he considers to be a dangerous doctrine. But because Ager writes against uniformitarianism, does not make him an ally of creationists. In fact, Ager does not like to be quoted out of context by creationists whom he considers to be unscientific in their thinking. The truth of the matter; he has a personal vendetta against them and he writes this terrible indictment against creationism:

"This [new catastrophism] is not the old-fashioned catastrophism of Noah's flood and huge conflagrations. I do not think the bible-oriented fundamentalists are worth honoring with an answer to their nonsense."

The New Catastrophism

Pp.xi – xix]

Remind me never to quote from Ager for the support of any of my beliefs in the catastrophic effects of the biblical Flood! To keep the information of this chapter on a down-to-earth level, I will address only the issue of Uniformity and the Flood of the Bible. There are many scientists who do not agree with new catastrophism. The doctrine of uniformitarianism is still a potent force in the tenets of geology. I see the two beliefs – Diluvialism and Uniformitarianism – as the two main opposing philosophies. They are in a conflicting position and I find it difficult to believe that some creationists are accepting Uniformity tenets and going so far as to deny the World-Wide Flood and to relegate the book of Genesis as mythological in content. The following clarification is necessary for my readers: I prefer the particular brand of Creationism that marches under the banner of Diluvialism rather than the banner of Catastrophism. Catastrophism is a word too vague. It insinuates that you can believe in geological catastrophes without adhering to the other beliefs of most Orthodox Creationists such as belief in the Universality of Noah's Flood, the Divine Inspiration of the Bible, the exact interpretation of the book of Genesis in describing the Formation of the Heavens and Earth, the Creation of the First-Parents, Adam and Eve, the Fall of Man described in Genesis, chapter 3 and his Redemption Promised in Genesis 3:15.

As a Diluvialist, I accept the historical and factual accounts recorded in Genesis 1-10. Creationists will be quoted who believe in the World-Wide Flood such as the biblical Flood (Noachian Flood). I apologize to those who are left out of the picture: neither falling under the category of *Orthodox Creationists* nor *Diluvialists* but I am not adept at being able to pinpoint and fine-tune the exceptions. Before making reference to our first source, I would like to say something about George McCready Price who passed to his rest Thursday, January 24, 1963, at the age of ninety-two. He reaffirmed and replanted the young-earth creationism doctrine through his writings. Many churches are indebted to him for his pioneer work in the field of geology: validating the truth of the literality of the book of Genesis, finding vast geological phenomena attesting to the reality of Noah's flood, adding to the knowledge that made common-sense of geology while demonstrating the illogical geology contained within the doctrine of uniformitarianism. In fact, Price sent about 500 copies (free of charge) of his book *Illogical Geology* to the leading scientists and theologians of his day.

True he lacked a Ph.D. degree in geology but he had a photographic mind and committed to memory all the necessary information that enabled him to write 25 books over a span of sixty years. During that course of time, he taught Latin, Greek, chemistry, English literature, physics, and biology to mention a few of his interests. I would have to belief that this agenda and schedule would not be among the capabilities of most men who earn a Ph.D. degree.

Price did a great deal of field-work in geology and frequented many libraries in his quest to learn about the *illogical geology* he dedicated his life to refute. I had the privilege of purchasing a copy of *Common-Sense Geology* from an old book store in Philadelphia, Pennsylvania. I was fifteen when I made the purchase for a dollar and seventy-five cents. My interest in geology started out as a flame in my nature and the fire has not gone out as I am presently seventy-one at the time of this writing. [The reason that I mentioned the purchase of the book is not because I have such a memory but rather, the price is written on the inside of the fly-leaf] Since the time of my reading *Common-Sense Geology*, I have devoured *The Story of the Fossils*, *Illogical Geology*, *Q.E.D.*, *The Phantom of Organic Evolution*, *Theories of Satanic Origin*, *The Fundamentals of Geology*, and *The New Geology*. I even read the debate that took place between Price and Joseph McCabe held at the Queen's Hall, Langham Place, London, in the 1920s. I realize that Price taught some things present creationists would no longer be in agreement. But this would be expected with the theories of a man born in 1870 compared with the present knowledge that is now available to creation scientists.

We are now ready to take up the task of determining the truth of the Geologic Column as it is viewed in the light of uniformitarianism and compared with the record of the fossils, contrasted with common-sense geology. Was *Archaeopteryx* in the proper time zone and a true intermediate between lizards and birds? Is the stratigraphic proof for a Darwinian origin of birds incontrovertible? Does bird evolution demonstrate a correct sequence in the geologic column? Does it all fall into place as evolutionists have predicted for bird transmutation? In fact, are the Geologic Column and its Zones of Time a matter of record or are these so-called facts just a matter of imagination and fantasy? It may very well be, Robert Bakker has no geologic column for supporting his argument. The column may not have any existence at all.

Price, throughout his teachings, used the metaphor of the coroner. He likened the geologist to the work of a coroner since they both judge the cause of death. The following quotes will serve as examples of his favorite illustration of those who are responsible for rendering a strict verdict of how species died. These men must be without prejudice and must decide according to the evidence.

In his book entitled *The Story of the Fossils* he writes:

"A CORONER has many criteria by which he may judge whether the body he is examining met with external physical violence or whether it shows signs of natural death. Similarly a geologist can usually judge whether the fossils he discovers were buried in some extraordinary manner or in the more orderly routine of everyday natural processes."

P.13

Again he writes of the coroner:

"If a coroner were to proclaim himself as a complete disbeliever in the possibility of a murder or a suicide, and declare that all deaths are due to 'natural causes,' such as sickness or old age, he would be hooted out of his official position without delay. We want the truth, the whole truth, and nothing but the truth in such cases."

Ibid.P.6

One more quotation from *Common-Sense Geology*:

"Geologists are only coroners holding an inquest on the earth, for the same principles hold true in the work of geologists as in that of coroners. And it is absurdly unscientific for us to decide by snap judgment in advance that the great geological changes of the past were like modern everyday processes if there is abundant telltale evidence that they must have been caused by some great world convulsion in the long ago."

P.27

I suppose that Price was tired of "snap judgments" when it came to scientists making decisions in favor of uniformitarianism and against the common-sense geology that should have had its physical events interpreted by way of catastrophism which was so obvious throughout the world. It was a "snap judgment" to believe in slow or deep time when the sedimentary nature of the rocks indicated fast deposits within a short bracket of time. There was "telltale" evidence that animals met their death in a violent and unnatural manner. It was a "snap judgment" to view them as dying from natural causes.

Uniformitarianism changed the hearts of many men and women: Charles Lyell and his book, *The Geological Evidence of Man* became more popular than the books of Moses, especially the first book Genesis that tells of man's creation, his fall, and his redemption. A "snap judgment" was made in believing that man was not created in the image of God but rather in the image of some primate.

The darkness of *The Origin of Species* overshadowed the light of the Bible. The light shone in the darkness but men understood it not. The prophecy of the apostle Peter recorded in his epistle (2 Peter 2:3-13) and condemning uniformitarianism, warning those who reject the Noachian Flood, admonishing people to remember that the world was once destroyed by water - was rejected. In place of repentance, came scoffing and derision of God's Word.

Even the words of Jesus which bespoke of the World-Wide effects of His second coming, using the World-Wide judgment of the Flood - was forgotten. Those that remembered His saying did not care: "But as the days of Noe were, so shall also the coming of the Son of man be. For as in the days that were before the flood they were eating and drinking, marrying and giving in marriage, until the day that Noe entered into the ark, and knew not until the flood came, and took them all away; so shall also the coming of the Son of man be."

Matthew 24:37-39

KJV

Many "snap judgments" that were made in the time of Price, are being made in our time and will be made in the last days when the earth will be purged by fire.

Let us now look at some of the evidences of flood action in the geologic column that tells a highly different story than what is told by evolutionary geologists. Let us consider the state and condition of the fossils, dinosaur graveyards, the high velocity of wave action and its role in sedimentary deposits and the various strata anomalies or irregularities discovered in rock formations.

(1) The State and Conditions of the Fossils

Apparently the verdict of Darwin's coroners has given a false report on the state and condition of the fossils. The geologic column is purported to show us living forms of life slowly winding their way up the column from past life until present life and all these animals were supposed to have died of natural causes (such as old age and sickness) and as victims of predators. The species are fossils and fossils are evidence of "past" life, so said the coroners of evolutionary geology.

Not only are fossils dead forms of past life but is admitted by many evolutionary scientists that such fossils require catastrophic burial. How can animals buried by an unusual event tell of evolutionary progression? There is nothing to designate they were not contemporaneous and that the geologic time column was not indicative of a single catastrophic event. In other words, the study of geology and paleontology does not show evolution of life over the span of millions of years. Rather, these sciences demonstrated the cataclysmic destruction of life in one age. If this is so and all the evidence points to the Great Flood, then the horizontal lines drawn through the earth and depicted in the geologic time zones, are illusions painted by the imagination of Uniformitarianists.

Henry M. Morris and John D. Morris write:

"Creationists have always been aware, of course, that uniformitarian geologists (even Lyell) have allowed for local catastrophes (Floods,

184

volcanic flows, earthquakes, etc.), but it is only recently (Shea and Hsu notwithstanding) that most geologists have been willing to consider such things as regional and even global catastrophes, as well as sudden global extinctions of many forms of plant and animal life. Creationists have been citing evidences of global catastrophism for years and at last secular, evolutionary geologists are beginning to recognize them too."

The Modern Creation Trilogy (Vol.2)

P.261

One of the supreme books on geology and recently prepared is *Footprints in the ASH*. I say this since the book is not lengthy, is easily read and understood, contains crucial information which undermines the tenets of Uniformitarianism, all information accompanied by beautiful photographs and was written by two outstanding creation scientists – John Morris and Steven A. Austin. John Morris earned his Ph.D. in Geological Engineering and Steven Austin received his doctorate in geology from Pennsylvania State University. Austin is one of those geologists who enjoys field-work and can't wait until he reports his findings. After reading this book, I was miffed when contemplating those individuals who still maintain the doctrine of uniformity is a thing to be cherished. Austin has cited many geological events as they occurred after the eruption of Mount St. Helens. I was amazed with the number of incidents that confirmed the catastrophic nature of geologic events and how they were responsible for shaping *physical geology* within a *short period of time*. These were geological incidents recorded in a diminutive time period and if this explosion had occurred in the distant past, there would have been blocks of time recorded by Uniformitarianists of the present moment, equal to many millions of years. I was especially fascinated by the reports on sedimentation and the photographs that accompanied the events. The scientists Morris and Austin wrote:

"Keep in mind that this is just the upper portion of a much thicker deposit. Geologists normally think that it takes excessively long periods of time to accumulate such thick sequences of sediments. In this picture, however, we see three episodes of rapid sedimentation recorded; each one took minutes to hours instead of long periods of time to form."

Footprints in the Ash

P.53

Austin autographed my personal copy of his book and added "Psalm 46." Verse eight in this Psalm reads:

"Come and see the works of the LORD, the desolations he has brought on the earth."

These words can be applied to the Mount St. Helens episode. God bids all scientists to observe His works which are "a teaching tool to the earth scientist." [Ibid.P.12]

And what can scientists of the uniformity persuasion learn? Simply this: the catastrophic happening at Mount St. Helens took place in a short time period. The results of these geological events conclusively demonstrate that geological formations can happen quickly and do not require long periods of time. We can compare such events with the entire geologic column and, through this analogy, can learn much about the rapid rate of catastrophism and its affect upon the earth.

The events recorded in the geologic column, if caused by a word-wide flood, could have produced far greater destruction than Mount St. Helens. Because each geological period is fraught with a catastrophic event and since *all periods are linked together by an essentially continuous deposition process* such as would be caused by the Great Flood, we can logically assume that these separate components were without a significant time break and therefore, we are justified in viewing the entire geologic column as being equal to the sum of its parts. This, as a large consequence, would erase the *Imaginatory Lines* separating long periods in the TIME SCALE. There is also geological evidence that the same basic patterns of nature mark each and every time zone. In other words, the entire Geologic Column has the same physical prototypes. That is: the same lithographic structures, the same form of minerals, etc. These are global phenomena, erasing the imaginary lines of the various time divisions.

CONCLUSION to the state and condition of the fossils. Because of this worldwide conformity only one conclusion can be formulated to all this: animals and plant fossils did not exist and die a few at a time, in a system of relays covering hundreds of millions of years. Because of fast geological events and since the sediments indicate signs of being laid down at one specific time, various time periods were essentially erased and thus the fossils were contemporaneous at the time they met their death and extinction.

(2) Dinosaur Graveyards

Certainly graveyards attest to the fact; animals and plants died of abnormal causes. What does it mater that a certain geologist brings in his coroner's report that tells us otherwise? Whatever is our view of fossils being presently formed, no one can say that fossilization is taking place on the same level as recorded in the past. There are many hundreds of fossil beds recorded throughout the world.

Since this book has mainly relied on dinosaurs, the dinosaur graveyards will serve as our best testimonials to the biblical Flood recorded in Genesis 6-9. Keep in mind, while we are discussing these graves, how difficult it is to discover large land animals being fossilized at this present time. Fossils of huge land creatures are practically nonexistent and when compared to bone quarries of similar sized animals of the past, set up a vivid contrast. Paleontologists find it difficult to explain this disparity.

Our first examples will be from the United States of America:

"The discovery of complete fossil skeletons is indeed extremely rare. The smaller *Coelophysis* is an astounding exception. At the Ghost Ranch in New Mexico there is a single quarry, where dozens upon dozens of complete articulated skeletons of *Coelophysis* were found side by side, and jumbled on top of each other. The partial and complete remains of hundreds of them have been removed, and yet the quarry has only been partially worked out. The total number of *Coelophysis* at this mass grave site has been estimated as representing several thousand individuals ….

"Many unknown factors may have been responsible for the great accumulation of *Coelophysis* in this particular quarry. Even more puzzling is the cause of their death. What natural catastrophe could have been responsible for the deaths of thousands of these small dinosaurs? Could fire or floods or disease or even accidental poisoning have been responsible? The exact cause remains a mystery."

Dinosaurs A Global View

Sylvia J. Czerkas

Stephen A. Czerkas

Pp.104-105

In analyzing this graveyard situation, no one is able to claim first hand observation of the transpired events which led up to the death of these "mighty reptiles." Since no one was there to see the "mysterious" occurrence, we must assume the role of the coroner in determining the cause for the demise of thousands of *Coelophysis* dinosaurs.

The Czerkas artist-paleontologists have pointed to a *natural* catastrophe. This is not unusual to my way of thinking. Most evolutionists, in their denial of the World-Wide Flood, would determine the quarry to be the result of a *natural* cause. For Czerkas, it is not the *type* of catastrophe that is a mystery; that has already been determined – it is a *natural* catastrophe. The only thing that remains a mystery is the *source* of the catastrophe. Czerkas writes, "What natural catastrophe could have been responsible for the deaths of thousand of these small dinosaurs? Could fire or floods or disease or even accidental poisoning have been responsible?" Czerkas recognizes the cause of these thousands of dinosaurs banding together was *not because they were gregarious* since there is evidence that they were *cannibalistic.*

Czerkas cites other dinosaurs, virtually indistinguishable from *Coelophysis*, found in South Africa, Europe, Asia and South America. In the South African quarry, two dozen individuals are jumbled together, in the remarkably small space of only a few meters in diameter. Czerkas *suggests* that *Coelophysis* stayed together in large groups.

The Czerkas conclusion to the New Mexico *Coelophysis* quarry is this:

- *Coelophysis* banding together in extremely large concentration; is suggestive of the nature of gregarious animals like herbivores. However, these *Coelophysis* meat eaters could end up on the dinner table of their fellow-feasters. [He infers that "gregarious" is not the proper connotation and thus, the reason for the concentration is not fully understood and remains a mystery]

- Death by way of NATURAL CATASTROPHE is implied to be the only certain feature in the Czerkas conclusion.

- The "unknown factors" responsible for the "great accumulation" of Coelophysis is puzzling but what is "even more puzzling is the cause of their death." The final admission is – THE EXACT CAUSE OF DEATH REMAINS A MYSTERY.

The Czerkas judgment is typical of the conclusions of dinosaurologists who have studied such graveyards that are found all over the world. These men, who are transfixed in the doctrine of evolution, without exception point to *a natural disaster* and almost always claim the quarry to be a *great mystery* and are virtually miff and bewildered in their coroner's explanation.

I would like to venture an opposing option. Since multitudes of dinosaurs have been wiped off the face of the earth with no satisfactory explanation for the demise and extinction of this mighty race and their disappearance remains inexplicable, NATURAL CAUSE cannot be the feasible answer. I would suggest that the only viable explanation is the ABNORMAL CAUSE – the Great Flood of the Bible (Genesis 6-9). These ancient graveyards cannot be accounted for through the weak and pathetic explanation of natural cause due to whatever catastrophe meets the Uniformitarianist's fancy.

My mind has a vivid picture of what happened way back then and the image produced in my thought processes is much different, I am certain, than what is produced in the thought pattern of the evolutionist. I see the floodgates of the heavens opening and the rain falling to the earth. I see lakes gradually forming on the plains of what would one day become New Mexico, driving the *Coelophysis* to seek higher ground as they *banded together to escape death*. Thousands of them scamper on top of each other to *escape the rising waters* as the flood is covering every inch of ground. The *Coelophysis* are covered quickly, waiting for that moment when their fossils would testify to the validity of the Great Flood. I, of course, was not at the scene when it took place. However, the panoramic view which I have of the specific and unnatural catastrophe is far more satisfactory to me than that evolutionary view which has considerably less facts for its backing and mostly lacks logic and reasonability.

The account of the *Coelophysis* skeletons and the countless thousand of fossilized bones around the world, tell us something. In a discussion on fossils, Alfred Rehwinkel writes:

"What wonders of a strange but perished world the fossils reveal? But as we examine them, whether they are found in America, Europe, Asia, or Australia, or any other place on the face of the earth, they tell one and the same story, and that is a sudden, wholesale destruction followed by an immediate burial. Only one force known to man is capable of accomplishing that, and that force is water. Hence we conclude that the fossils found in every part of the world constitute convincing evidence for the Biblical Flood."

The Flood

Alfred M. Rehwinkel

P.237

We go eastward from Ghost Ranch in New Mexico to the Cleveland-Lloyd quarry in central Utah. You must realize that I am giving you but a small sampling of these dinosaur quarries. The fossils found in this quarry contain a large number of vicious *Allosaurus* – cruel *meat-eaters*. Czerkas informs us:

"Also, with only few exceptions, the bones are totally disarticulated and randomly scattered ....

"Exactly how and why they all came to rest in this quarry is quite mystifying and uncertain. It is possible that it was a natural trap for predatory dinosaurs. Tiny shells of snail-like gastropods and other fossils suggest that this was once as area of shallow, calm, slow moving water."

Dinosaurs A Global View

Sylvia J. Czerkas

Stephen A. Czerkas

P.151

For Czerkas, this quarry is another mystery engulfed in uncertainty. It is possible, as Czerkas suggests, the quarry was a "natural trap" but not very likely. No tar was the ingredient at Cleveland-Lloyd – only mud. Further in his book; Czerkas proposes that the mud may have turned into "a deadly quicksand bog" and "Many of the bones in the quarry show signs of predation, and some even show clean breaks as if trampled and broken either soon after or during the last moments of life."

Ibid. 151

Some bones showed "clean breaks as if trampled and broken either soon after or during the last moments of life." Being trampled upon in the last moments of life indicates a strong possibility that they may have been fleeing from the rising water of the Great Flood.

The tiny shells and snail-like gastropods could indicate that the quarry was once the sight of shallow, calm, and slow-moving water and the mud it left behind served as a trap for the scavenging *Allosaurus* But there is also the possibility; the *Allosaurus* came to rest in the quarry as they were washed in from some other environment. Even evolutionists would admit that animals buried by flooding, are not always covered *in situ.* That is, in the particular environment or zonation in which they lived. To the creationist, the possibility exists; fast-moving currents of water carried the *Allosaurus* a long distance before depositing them or dropping them off in another environment.

The Uniformitarianist, of course, looks to see evidences of a natural catastrophe without ever considering the abnormal activities of the catastrophic nature of the Word-Wide Flood. Czerkas describes the bones, and with few exceptions, are totally disarticulated and randomly scattered. I do not know weather or not mud has the same effect on fossil bones as does tar! In the La Brea pits there is a term called "pit wear." Tar can move bones around and produce grooves or cuts. But even if mud can manufacture the same phenomenon causing grooves or cuts, I have a serious doubt; it can bring about the dismembering and total disarticulation of bones.

Again, I sense the evidence points to a more potent and dynamic catastrophe rather than to the localized theory of "entrapment." The space of this book will not allow further discussion on the quarries at the Dinosaur National Monument, Dry Mesa, Garden Park in Colorado, etc. All of these quarries tell the same story – the burial of tons of fossils by water, the result of a raging and unnatural flood and the "wholesale destruction followed by an immediate burial" (Rehwinkel). I feel that the geological and paleotological implications of all this points to and verifies the Great Flood of the Bible.

I realize that many men of this world, disparage the Flood and the Christian Bible but the scriptures are filled with the words and writings of the prophets who wrote and proclaimed the truth about Creation, the Flood, and Redemption; the disciples who could not deny the prophets and preached the same message; Jesus who lived and verified the truth of the prophets and the disciples whom He had called into service and finally: as Creation testified to him as Creator, the Flood established Him as Judge of this world, the Cross became the symbol of the life He surrendered for us as our Redeemer.

Time for one more episode: only this time we will go across the sea to a Belgian coal mine for our final dinosaur mass burial. In 1878, in the village of Bernissart in Belgium, Coal Miners discovered bones of the dinosaur, Iguanodon – a large plant-eating dinosaur that grew up to 30 feet long and weighed close to five tons. Actually, there was a concentration of *Iguanodon* skeletons discovered a thousand feet below ground. Officials of The Royal Museum of Natural History in Belgium were informed and paleontologists were sent to dig out the dinosaurs. The project took about three years. The skeletons of Bernissart were articulated, that is, complete specimens. About 39 skeletons were collected until flooding halted the project.

There is a variety of theories as to how these creatures died and were entombed. How does one account for the remarkable accumulation of skeletons in one area of interment? Let us briefly review the judgments of the paleontologists who were serving as *detective-coroners* of that time:

1) The dinosaurs did not die at one specific time and there appear to be four different species. Their bodies were covered by the sediments of a marsh or lake. Some were found in mudstone and others in sandstone at different levels. There was also coalified plant material in the dinosaur graveyard.

2) There was a deep fissure ["Cran du midi"] in the earth and it was deemed quite possible, the *Iguanodons* tumbled to their death while they were being driven over the cliff by a predator or two, *Megalosaurus*. Gish writes on this theory:

"How to account for this amazing fossil graveyard of *Iguanodons* deep in a coal mine has posed a formidable challenge to evolutionary geologists. Some have suggested that at one time there was a deep fissure in the earth into which these *Iguanodons* tumbled to their death. It seems strange, however, how all other animals managed to elude this trap. Only a few lonely voices have suggested that these creatures and the coalified plant material in which they were entombed had been swept there and buried by a vast aqueous catastrophe."

EVOLUTION: the fossils STILL say NO!

P.116

[Pittack's Note: There were five turtles, 3,000 fish (sixteen species), and five kinds of crocodiles as well as one salamander but they were all aqueous creatures and it is *still remarkable* that no large, land animals were found within the trap. Perhaps *there was a lake* at the bottom of the ravine which served as a habitat for some of these animals. The remaining animals, along with the *Iguanodons*, could have been washed into the ravine by the Great Flood]

3) The animals perished suddenly and collectively and since there were no juveniles in the lot, old individuals went to the ravine to die and were covered by floods that probably occurred regularly. Few paleontologists accept this theory while others view it as a joke.

A number of things are evident in these theories. Firstly, in order for fossilization to take place, there had to be immediate burial. Secondly, the clay and sandstone levels in which the *Iguanodons* were entombed were water deposited. In fact, all sedimentary rock is formed by solid fragmental material, transported mainly by water, and deposited in layers. Thirdly, at least one of these three theories associates *flooding* with burial.

Creationists would suggest, and with good reason, that the I*guanodons* and the coalified plant material were swept into the ravine, buried by a vast aqueous catastrophe, and entombed. There is no reason to assume that the dinosaurs died one at a time although they are found in different levels of clay and sandstone. Such levels of strata could have been laid down quickly by the Great Flood and deposited from different areas. This would account for the various *Iguanodon* species that were washed from their own habitats. While evolutionists compete one with another for the plausibility of their theories, creationists have as much right to present the logic and credibility of their understanding of the burial of the Belgium dinosaurs.

CONCLUSION to dinosaur graveyards: Much evidence has been presented to show evolutionary theories are often contradictory. This fact should tell us something about the geological and paleotological decisions and judgments which serve as the basis for each theory regarding both the local and the final eradication of mighty reptiles. There are 80 to a 100 extinction theories in circulation and it is impossible to declare any one scientist as the arbiter of truth. On the other hand, creationists who believe in the biblical Flood are convinced their affirmations are *scientifically grounded* since they are supported by *empirical facts* within the geological and paleotological systems viewed from within the *creation-model* [See Pittack's book,

Dinosaurs were obliterated from off the face of the globe and their kind is unknown except for what has been preserved in the fossil record. Evolutionists consider dinosaur extinction to be a mystery. Their scientists still find it difficult to explain why dinosaurs perished simultaneously over all the earth and how they were massed together in large cemeteries.

Creationists do not find dinosaur extinction to be a mystery. They believe the answers and solutions are provided and contained within the biblical account of the Flood Catastrophe. It is reasonable to conclude that water is the only agent that could bring about all the necessary conditions for dinosaur destruction and burial on so wide a scale. The evidence shows: the fossils discovered in graveyards throughout the world have a common source for their extinction – THE GREAT FLOOD. This being true, the sedimentary nature of the Geologic Column produced by this same catastrophic force; the Time Divisions marked by periods of the Mesozoic become purely imaginatory.

The horizontal, phantom lines running through the surface of the earth are "erased" and become non-existent. Since the dinosaur graveyards show a massive catastrophism throughout the Triassic, Jurassic, and Cretaceous levels, at least the so-called Mesozoic era is erased. The four other great "eras" in their arbitrary organization are also erased since massive catastrophism is responsible for the major extinctions of most other animals buried throughout the Geologic Column. Since there are mass extinctions that took place at every level of the geologic column, we can assume that ONE EVENT caused animal and plant death. Because the whole column is equal to the sum of its parts, the ERASURE of the horizontal time levels is duly noted at EVERY TIME DIVISION THROUGHOUT THE GEOLOGIC COLUMN.

> (3) The High Velocity of Wave Action and its Role in Sedimentary Deposits

The rapid formation of sedimentary rock systems attests to the catastrophical force of water. Anywhere we look in the world, the sedimentary rocks are nature's object lessons of extensive and energetic water action. The following examples could be multiplied a thousand times with like-statements from the geological literature.

Harold W. Clark writes:

"If we travel over the Colorado Plateau .... We would find extensive cross-bedded shales and sandstones .... Some of these beds are comparatively thin, sometimes from ten to a hundred feet thick, but spread out quite regularly over a vast area of 100,000 square miles or more. How such deposits could have taken place without violent and widespread waves of water is impossible to imagine ....

"In South Africa are to be seen great masses of schist and granites which are described as having been spread out and sorted by immense currents of water. They have been likened to giant placers which worked over vast quantities of material. Certainly there is nothing of a normal or uniformitarian nature in this."

Fossils, Flood, and Fire

P.32

Surely the above examples of fast action deposits offer evidence that challenge the postulation of uniformitarianism and supports the Theory of Diluvialism. Because rapid rock formations are formed catastrophically and by fast moving water throughout the strata of each time period in the geologic column, how can we fail to deduce: the entire column is one mass unit. Henry Morris and John Morris offer this logical and common-sense observation:

"Now, if every unit in the geologic column was produced by at least a local catastrophe, and all the units are connected through an essentially continuous deposition process, then the entire series must represent merely different local components of the same worldwide cataclysm. The whole is the sum of its parts."

The Modern Creation Trilogy

SCIENCE AND CREATION (Volume 2)

P.281

Again, it is evident the Doctrine of Uniformity does not fit into the context of the present being able to tell us the whole truth about geological events of the past. What have we learned thus far in this chapter?

From the state and condition of the fossils, we learned that plants and animals did not die a few at a time but they were contemporaneous at the time of their death and extinction. Thus, *various time zones are essentially erased.* From dinosaur graveyards, we learned that the only possible explanation for the extinction and demise of the dinosaurs was the World-Wide Flood. Thus, the so-called Mesozoic Era was essentially erased and because the same catastrophic event was responsible for killing other plants and animals besides dinosaurs; the so-called Paleozoic Era *is also essentially erased* and *most* of the Cenozoic Era. I say "most" because "it is not easy to draw a line between the last Flood deposits and the post-Flood sediments" (Clark). From the high velocity of water action in depositing the sediments, we learned that such action could be caused only by the World-Wide Flood. The forces of water currents are seen at every level in the \Geologic Column. Sometimes they can be attributed to local flooding but other times they can be viewed only as stemming from the horrific and terrible force from a grandiose Flood that must be deemed as very unnatural and abnormal, even to the point of being considered a paranormal event. Since the entire sedimentary Geologic Column is a unit all over the world and without a break in time, *the various time zones are essentially erased.*

Let us move on to the significance of the different rock anomalies.

(4) The Various Strata Anomalies or Irregularities Discovered in
   Rock Formations

So far we have learned to distrust the scientific value of time zones in the Geologic Column and the reasons why. Rock anomalies or deposits of rocks in the sedimentary column, which are contrary to the evolutionary order of the fossil record, will simply add more evidence to enhance the mistrust of time zones.

I realize the Geologic Time Scale is considered to be the "sacred cow" to most evolutionists but it is time to give up "the vaunted supremacy of the Geologic Column" (Woodmorappe) and stop believing "From the beginning of its history, the earth has maintained a detailed geological journal ..." (Mc Laughlin).

John Woodmorappe has written a book entitled "Studies in Flood Geology" which is a compilation of research studies supporting Creation and the Flood. His studies are technical but, for the most part, understandable. His reports establish the fallacy of McLaughlin's above statement.

The following geological phenomena can be problematic to the evolutionist. Such phenomena are called anomalies. Only a few will be considered.

- PARACONFORMITIES – Here is the definition given by the "Dictionary of Geological Terms" edited by Robert L Bates and Julia A. Jackson: "An obscure or uncertain unconformity in which no erosion surface is discernible or in which the contact is a simple bedding plane, and in which the beds above and below the break are parallel" (P.368).

This formation is only "obscure" and "uncertain" to evolutionists. "Unconformity" is a word used by evolutionary geologists. In simple terms: the rock formation does *not confirm* to the beliefs of evolutionists. The formation upsets the "apple-cart" so to speak. In this case, the rocks tell a different story than that which is propagated by the evolutionist who informs the public the fossils is in proper order in the Geologic Column. How can the fossils show a proper order if some of the order is MISSING.

A Paraconformity is a more recent term than is "deceptive conformity." Using the old term, placed evolutionists in a bad light since it made creationists think that the only people who were truly being "deceived" were the evolutionist themselves. Common-sense geology enables us to *accept* what we see. This is the more empirical way of observing and accepting nature for what it is and not the way we would like it to be. The evolutionist *observes* the geologic column but he is *unable to accept* what he sees. But what is it that he sees? He sees a common phenomenon in geology wherein a younger bed is resting conformably upon a much older bed. Or, in some cases, the "old" formation rests conformably on the "young" formation. These cases would serve as examples of "reversed ages."

And what does this indicate? – Many intervening geological ages have vanished, entirely missing. There is a strange thing about all this: there is no evidence of sedimentary discontinuity such as erosion. The definition at the beginning of this section bears this out. If the two formations under consideration are parallel and there is NO EROSION; it is *Illogical Geology* to say, "There must be a missing section." Why should the *Coroner* pencil this alleged fact into his note book when there is no evidence to support such an idea?

In truth, the strata are not missing but appear to be missing to the evolutionist who banks on the so-called missing *ages of time* to confirm his belief about the order of life in the fossil record. He depends on the sequential order of the fossils and when there are various time zones missing: he labels this "obscure" discovery, a Paraconformity ("deceptive conformity").

[Pittack's Note: The public is supposed to accept all this *as true and factual science.* "What do we know about geology and Paraconformity? Just tell us that evolution is true and we will believe it", they say. The geologist steeped in evolution, reports to the public *what he believes and not necessarily what* is seen. The public is often deceived into believing in evolution because people give credence to the following syllogism:

Major Premise:	I believe in true science.
Minor Premise:	Geology and paleontology are true sciences, which teach evolution.
Conclusion:	Therefore, I believe in evolution.

What a deceptive syllogism! – The minor premise is incorrect. Geologists and Paleontologists cannot always be relied upon to speak absolute truth simply because their specialized fields often involve theory and speculation, especially when it comes to the evolutionary philosophy.

Here is the bottom line: There are no missing strata. The strata do not exist except in the minds and imagination of evolutionists who doubt what they see in nature but do not doubt what they see in their imagination. Again, the imaginatory time levels are erased in the Geologic Column but in this particular case: *how can you erase something that has no existence in the first place?*

- OUT OF ORDER STRATA - All the evidences that have been mentioned so far, hardly gives us room to believe in such a thing as the stages of life throughout the eras of the Geologic Column, telling us when new species came upon the stage of existence and when they became extinct and new species took their place. If anything, we should be convinced by now, the geologic ages are quickly vanishing right before our eyes.

We have already familiarized ourselves with two stratigraphic anomalies - 1) Young beds on top of old beds; Old beds on top of young beds and now 2) "Out of order" strata.

Henry M. and John D. Morris write:

"There is really nothing in either the physical or paleotological components of the rocks that would necessarily require identification of rocks as belonging to different ages. In addition to that very basic fact, there also exist many anomalous deposits in the sedimentary column that make it still more difficult to believe in the standard geologic ages. One of these is the frequent occurrence of fossils that are found in deposits that contradict the 'official' evolutionary order. This phenomenon is more or less brushed off with the term 'stratigraphic disorder.'"

The Modern Creation Trilogy

SCIENCE & CREATION

P.300

To every professional "stratigrapher" who studies earth's strata and records the facts, you would think when fossils are found in "disorder." This would appear as a recorded fact in the geological reports. This "stratigraphic disorder" is a very common phenomenon. However, it is not well documented. I wonder why! Could it be that such discoveries are *not in keeping* with the "official" order of fossils that the omission of facts is a sure sign of *scientific murder* as one evolutionist calls it, that it is a way of *escaping the truth* of non-sequential order, a way of *escaping problems*, which evolutionists do not care to discuss.

When men play around with scientific facts how can they, in turn, say of creationists: they are dim-witted, not worthy of a response on any issue, their so-called degrees do not entitle them to claim the title of "scientist." In fact, "creation-scientist" is the biggest paradox on the face of the earth, so say the evolutionists.

I was greatly surprised by this candid statement:

"...However, The widespread occurrence of anomalies in dated sections suggests that disorder should be taken seriously by paleobiologists and stratigraphers working at fine stratigraphic scales."

Alan H. Cutler and Karl W. Plessa, "Fossils out of Sequence: Computer Simulations and Strategies for Dealing with Stratigraphic Disorder," *Palaios*, vol.5 (June, 1990), p.227.

P.301

What does it matter that the facts of science should be "taken seriously"? Just so long as evolution is believed to be the supreme philosophy and the public bows down to it as the Babylonians bowed to the golden image of Nebuchadnezzar's day.

There is no fixed time line in the Geologic Column, no sequential order of past life, no natural events to account for the demise of dinosaurs, no geological phenomenon to account for the distribution and heavy volume of deposited sediments, and no fossilization in the present time that will ever be on par with or be equal to the extensive fossilization of animals of the past..

At the beginning of this chapter, I informed my readers the reason for its existence was the statement of the dinosaurologist, Robert T. Bakker printed in his book, "The Dinosaur Heresies" –

"The stratigraphic proof for a Darwinian origin of birds appeared incontrovertible – the rocks preserved the stages of development in the exactly proper sequence through time. Any impartial observer might conclude that if God had really created birds, he must have been going out of his way to fool humanity into believing in evolution."

P.303

This chapter was designed to first of all confirm that the rocks and strata are not "incontrovertible" proofs for Darwin's origin of birds. The evidence presented thus far would make Bakker's argument to be inconsequential since the entire basis for his case has been removed. The complete Geologic Column has vanished with all its imaginatory zones of time. This chapter had a second purpose: to invalidate the thought that God fooled "humanity into believing in evolution." Humanity fooled itself by turning away from the Doctrine of Creation made known in the *special revelation of the Bible* and by corrupting and misreading *God's general revelation contained within nature.*

But for the sake of pursuing Bakker's contention one step further, it will be necessary to "restore" the Geologic Column for just a few moments. The truth of the *Archaeopteryx* paleontological position in the geological strata will now be taken up in order to ascertain the validity of Bakker's claim that *Archaeopteryx* is in the proper time period between reptiles and birds.

This final issue concerning the *Archaeopteryx* will now be addressed. Evolutionists besides Bakker have long believed one of the major points for accepting *Archaeopteryx* as a transitional form, is its place in the fossil record. That is, it appears after the so called evolutionary development of reptiles but prior to the development of modern birds. If this is true then it would be logical (not necessarily scientific) to conclude *Archaeopteryx* – with its reptilian and Avian characteristics – could possibly be an intermediary animal or a missing link between reptile and bird.

The truth of the *Archaeopteryx* paleotological position in the geological strata will be taken up in order to ascertain the validity of Bakker's claim that *Archaeopteryx* is in the proper time period between reptiles and birds. Francis Hitching is well known as a first-class science writer. Hitching is also recognized for his work on documentaries that appeared on television and dealt with inexplicable natural phenomena. The following statements demonstrate how he out-foxed Darwinists in his response to the *Archaeopteryx* fossil:

"But where are the fossils showing how fishes evolved into amphibians? Or how reptiles evolved into mammals? Or reptiles into birds? Missing, all missing, says Hitching.

'Not so!' cry the Darwinists, indignantly. 'Look at *Archaeopteryx*. That is half way between a reptile and a bird.'

Hitching is ready for them .... *Archaeopteryx* was formerly believed to have lived long before birds appeared, this is not so. A fossil was found in Colorado in 1977 of a true bird which could not have been descended from *Archaeopteryx*, because it lived at the same time."

Creation and Evolution

Alan Hayward

P.43

[Pittack's Note: Actually the discovery of the Coloradoan bird was older than the above report written by Hayward. This undisputed bird predated *Archaeopteryx*, according to the alleged time zone, by 60-million-years]

James A. Jensen, a geologist from Bringham Young University, turned bird evolution on its head that famous day in Dry Mesa quarry. The bird he discovered in the Lower Jurassic strata of Western Colorado refuted the concept that *Archaeopteryx* predates the general arrival of birds by millions of years.

Since a true bird was found in older strata than was *Archaeopteryx*, the later could not possibly be a progenitor of modern birds. It appears that Bakker is absolutely incorrect in telling the public that *Archaeopteryx* is "in the proper time period." The fossil IS NOT IN THE PROPER TIME PERIOD. Therefore, *Archaeopteryx* is NO INTERMEDIATE BETWEEN REPTILES AND BIRDS. It becomes apparent that the ancestors of flying birds must be traced back way beyond the time period in which *Archaeopteryx* lived – at least 60 million years.

There is a report that places the bird kind all the way back to the first dinosaur. Sankar Chatterjee and his two co-workers at Texas Tech University were fossil-collecting in the bluffs outside the town of Post. The rocks in this badland belonged to the Late Triassic Dockum Group and, according to geologists; date from the period about 225 million years ago. Chatterjee, in 1983, discovered a group of delicate white bones which he later cleaned and pieced together only to discover the anatomical characteristics were typical of a bird. The small bumps on the bones suggesting feather insertions and only one hole in the skull were strong indicators pointing to bird anatomy. You can imagine the excitement felt by Chatterjee as he realized this bird predated *Archaeopteryx* by 80 million years. This bird was named *Protoavis* or "ancestral bird" and Chatterjee considers the skull to be, in fact, to be more birdlike than *Archaeopteryx*.

Duane Gish comments in *EVOLUTION: the fossils STILL say NO!* :

"If Chatterjee's analysis is correct, then obviously neither dinosaurs nor *Archaeopteryx* could be the ancestral to birds. Furthermore, if birds really did evolve from reptiles of some sort, then a bird 75 million years older than *Archaeopteryx*, or 225 million years old, should be extremely reptilian. Chatterjee's *Protoavis*, according to Chatterjee, is just the opposite, even more bird-like than *Archaeopteryx*."

P.137

Whatever Bakker and other evolutionists claim for the so-called correct location of the *Archaeopteryx* fossil in the Geologic Column has been controverted by the evidence. Birds have not only been found at the same time level as the *Archaeopteryx* but in the *same time level of the reptiles*. These facts offer powerful testimony against the evolutionary theory: the *Archaeopteryx* was a *transition* between reptiles and birds. *Archaeopteryx* could not have been the "first bird" as Bakker and other evolutionists have claimed.

Robert Bakker writes that *Archaeopteryx* was a transition between reptiles and birds. BUT THE EVIDENCE SHOWS that *Archaeopteryx* appeared suddenly in the fossil record with no transitions leading up to it. This fits the creation model: *Archaeopteryx* was a created bird and formed with all the necessary equipment for flight in place and at one time.

Robert Bakker believes that *Archaeopteryx* provides insight into the evolutionary history of birds. BUT THE EVIDENCE SHOWS that it demonstrates no transitional features for the gradual development of flight from reptile to bird. All its attributes were fully developed and functional. Its wings and feathers were complete and perfect.

Robert Bakker informs the public that *Archaeopteryx* was discovered at the proper geological time period between reptiles and birds. BUT THE EVIDENCE SHOWS that James A. Jensen discovered a bird that was 60 million years older than *Archaeopteryx* and Chatterjee found a bird 80 million years older, that went back to the time of the first dinosaur.

Robert Bakker claims the Darwinian belief, regarding the evolution of birds, has been established throughout the Geologic Time Scale. BUT THE EVIDENCE SHOWS that the so-called time divisions are products of man's fertile imagination. The column has vanished in the following ways:

(1) The *state and condition of the fossils* verify they did not exist and die a few at a time over millions of years. Because of fast geological events, the sediments were laid down all at once. Animals met their death at the same time and *thus, the vast time periods were essentially erased.*

(2) The *dinosaur graveyards* indicate a common source for the extinction of these "mighty reptiles" – the Great Flood. Therefore, the *Mesozoic era has been erased* and since mass extinctions have buried most of the other animals throughout the Geologic Column in the same manner, we can safely say: the Flood was ONE EVENT and

essentially *erased* *the* *imaginatory,* *horizontal* *time* *levels* *of* the Paleozoic and most of the Cenozoic.

(3) The *high velocity of water in depositing the vast amount of sediments* could have been caused only by the World-Wide-Flood. Since the entire sedimentary, geologic column is a unit all over the world and without a break in time, *the various time zones are essentially erased.*

(4) The *various strata anomalies* discovered in rock formations have virtually destroyed the tenet: "From the beginning its history, the earth has maintained a detailed geological journal." *Anomalies* have proved the fallacy of this statement by *erasing the entire geologic column.*

This chapter has been lengthy but I have attempted to lead you through a problematic study of the evolutionary belief: *Archaeopteryx* has a place in the Geologic Time Scale. I am confident that you have understood the time scale is not only non-existent but even if it did exist, *Archaeopteryx* would not be in the appropriate order between reptiles and birds. Having written all this, presenting my case for creationism and in opposition to evolution, there remains a very important matter that needs to be addressed: the fossils, throughout the strata, often follow a particular order which seems to confirm the evolutionary teaching of slow progression of species and the Darwinian doctrine of transmutation. How does the creationist account for this phenomenon? This matter will be discussed in Appendix 1.

Towards the end of chapter seven, I informed you that we (Collectively speaking) have endeavored to bring out certain facts:

- The discovery of *Archaeopteryx*.

- The *Archaeopteryx* fossils are genuine.

- *Archaeopteryx* is not the "perfect" or missing link between reptiles and birds.

- Creationists are not unreasonable in asking for fossil transitions.

- Biochemistry and Cladistic Analysis do not solve the problem of missing links.

- *Archaeopteryx* is not a dinosaur nor is it a link between reptiles and birds.

- *Archaeopteryx* is not a reptile and no less a bird in spite of its reptilian characters.

At the close of this book, other facts have been added to your repertoire:

* The impossibility of feather evolution.

- Feathers make *Archaeopteryx* a full-pledged bird in its own right.

- Feathered dinosaurs do not truly exist.

- The "ground upward" theory and the "tree downward" theory are not sufficient in explaining bird flight in the *Archaeopteryx*.

- *Archaeopteryx* is not in the proper time zone and correct order between reptiles and birds. In fact, the Geologic Time Scale is imaginatory.

APPENDIX 1

IS CORRECT SEQUENCE OF FOSSILS A PROOF FOR EVOLUTION?

"The waters rose and covered the mountains to a depth of more then twenty feet. Every living thing that moved on the earth perished – birds, livestock, wild animals, all the creatures that swarm over the earth, and all mankind. Everything on dry land that had the breath of life in its nostrils died."

Genesis 7:20-22

NIV

With all the creationistic talk about non-sequential order of the fossils, there is *sometimes fossil order.* If there wasn't, then it would be incongruous for creationists to talk about fossil disorder as one of the geological *anomalies.* Just as the evolutionist must own up to the fact that there is such a phenomenon as "stratigraphic disorder"; so must the creationist own up to the fact that there is such a phenomenon as "stratigraphic order."

Creationists write about the *supposed order of the fossils.* "Supposed" is a literary device, in this case, to indicate that there is NO ORDER of the fossils. If this is considered to be a truism, then I cannot – as a fellow creationist – place myself in agreement.

However, I am in total agreement with the following facts:

- No fossil arrangement ever demonstrates exact "sequential order" since each local column seems to have its own variations from other columns.

- No fossil arrangement can ever prove that the various animals were not contemporaneous at the time of their death.

- No fossil arrangement can ever prove that plants or animals are separated in the rocks in which they were buried, by long periods of time or ages. [See Pittack's book, *Difficult Questions on Dinosaurs* and QUESTIONS ON DATING (DEEP TIME)]

- No fossil arrangement can ever prove evolution.

Some creationists would claim that each local column never comprises more than a minute percentage of the total column. However, practically the whole geological formation of rock is found in the canyons of the Wind River Mountains in Wyoming. Such an example shows the Pre-Cambrian level all the way up through the Pleistocene level. The following animal fossils are found:

Tertiary	Mammals
Cretaceous	
Jurassic	Reptiles
Triassic	
Permian	
	Amphibians
Pennsylvanian	
Mississippian	Largely Marine
Devonian	Fish and Marine life
Silurian	
Ordovician	Marine and Sea Bottom life
Cambrian	

[See *Fossils, Flood, and Fire*, Chapter six – "Ancient Life Zones", Pp.51-64, by Harold W. Clark]

How do creationists explain the above "fossil-rock sequence" in Wyoming? It matters not that the Wyoming example coupled with other examples are few and far between; they still have to be accounted for and explained by creationists.

The last chapter of this book places in plain words why uniformity was not the answer to the so-called time zones in the Geologic Column. Rather, the violent and widespread waves; immense currents of water, enormous amounts of sediments deposited, could be accounted for only by a World Wide Flood. The Biblical Flood was the answer to the Geologic Column but in "area" rather than in "time." Most creationists see the Column not as representing geological "ages" but "stages" in flood action. The animals and plants were buried by water in their environments. They died in the sediments representing their "Zonal Conditions."

Why do the fossils, in some instances, follow a *stratigraphic order*? A perfectly good answer is given by Henry and John Morris in *The Modern Creation Trilogy*:

"As a general rule, it is obvious that – other things being equal – such a global cataclysm would tend to deposit organisms in ecological burial zones corresponding to the ecological life zones where they were living when caught up by the global flooding. Thus, organisms living at lower elevations (deep-sea marine invertebrates) would usually be buried at lower elevations, while organisms living at higher elevations (bears, birds, etc.) would tend to be deposited at higher elevations.

"Another factor necessary to consider is that of mobility. Animals that are capable of running, swimming climbing, or flying can escape burial longer than others. One would, for example, expect to find few fossils of birds or human beings for this reason. Even though finally overtaken and drowned in the rising floodwaters they would be much less likely to be trapped in the sediments and preserved as fossils than other less mobile animals."

Pp.306-307, volume two

These explanations are logical and simple. But the reasons are not as simple as to render them having no value or not to allow room for further evaluation.

There are of course exceptions why the Flood would not bury animals and plants in life zones or natural habitats. The best run down of examples is given in *Fossils, Flood, and Fire*:

"Some cautions [to the theory of 'age' of time being replaced by a 'stage' of Flood action] must be observed, however, and we shall point out a few.

"Obviously in a series of deposits produced by cataclysmic action, we could not expect to have the correlation between the sequence in the rocks and the original positions of the living types absolutely exact. The violent wave action and the action of strong currents would, naturally, produce some degree of mixing of types.

"Some types that might have lived in certain habitats may have been able, because of their mobility, to escape to higher levels before they were overwhelmed. Thus their relative position might not be correctly represented. ...."

"We must not suppose that all types of life would be buried *in situ*, that is, right where they lived. The violent currents .... Would carry some forms for a long distance before dropping them ...."

Harold W. Clark

Pp.58-59

In general, what sequences there are and what sequence there is, are not due to the transmutation of species which developed over long ages of time but is rather due to the burial of life forms in their natural habitats or ancient zones.

APPENDIX II
A SPIDER'S WEB

RBP

APPENDIX 2

"Such is the destiny of all who forget God; so perishes the hope of the godless. What he trusts in is fragile; what he relies on is a spider's web. He leans on his web, but it gives way; he clings to it but it does not hold."

Job 8:13-15

NIV

Mark Isaak, in his book, *The Counter-Creationism Handbook* has written an entire section against intelligent design charging that such a theory is contrary to common sense. As part of his argument he writes:

"Spider webs apparently meet the standard of specified complexity, which implies that spiders are intelligent. One could instead claim that the complexity was designed into the spider and its abilities. But if that claim is made, one might just as well claim that the spider's designer was not intelligent but was intelligently designed, or maybe it was the spider's designer's designer that was intelligent. Thus, either spiders are intelligent, or intelligent design theory reduces to a weak Deism where all design might have entered into the universe only once at the beginning, or terms like "specified complexity' have no useful definition."

P.250

God forbid! But if I should ever commit an offence serious enough to merit a court procedure, without a doubt, I would have Isaak plead my case as the defense attorney. By the time he finished with his closing remarks: the jury, judge, and prosecuting attorneys would be so confused and discombobulated, there would be no recourse except to dismiss the case. [Most creationists are not deists. They believe in God not only as the intelligent Creator of the world but He is also the Controller of natural laws and guides the world through His intervention]

In the year 2004, just before Christmas, there appeared on the web site, a piece similar in "design" to Isaak's argument. Bob Fink claimed that no intelligence is behind this "marvel" of spider webs. The comments by Mr. Fink appeared in the site entitled, "The Holy Bible – Biblical Contradictions & fallacies, King James Version" (http://www. webster.sk. ca/greewich/bible-a.htm).

The following is my response:

Firstly, Fink points out God's intelligence must be divorced from the situation because the God of creation does not exist except in the minds of religious creationists.

Secondly, Fink claims the spider is mentally "asleep" and even "comatose" when it comes to making a web. Fink names the spider "Charlotte" after E. B. White's perennial best-selling novel. But instead of viewing the story as a tale of friendship and salvation, Fink disparages the beauty and moral of the story by calling the spider "dumb Charlotte."

In fact, Fink goes so far as to say, "The web is one example of a marvel made by a moron ..." I am spellbound with Fink's deep regard and love for nature in calling the spider "poor retard." Fink confesses to a daily wrecking of the spiders web but he did this only to prove a point. He proved the spider to lack intelligence by her failure to build a new web far away from Fink's daily route. He puts words into Charlotte's mouth which she should have said but didn't since she lacked intelligence: "I'll spin webs far away from his (Bob Fink's) stupid path." Perhaps the "stupid path" best describes this essay intended to be written in behalf of the best interest of science and the worst interest of creationism!

Fink gives some *brilliant* examples of marvels made by other morons in the world of nature. He attempted to write in a poetic style but in doing so, obscured the main point of his essay: the thinking powers of animals which Fink regards as "asleep" and "comatose." :

"If the web is one example of a marvel made by a moron, can't there be others? What do silkworms 'know' of silken beauty? [He probably means what do silkworms "know" about the mechanics of weaving silk and not the abstract notion of beauty] Or song birds of lilting music? [I suppose lilting is a typographical error for *lifting* and he means what do songbirds know about their own songs and their relationship to mating, alarm calling, etc.; not to the emotional effects of music] Or seeds of the geometric wonder and colorful designs of flowers? [The comparisons and association really break down at this point. What do seeds have to do with thinking? And, be careful Mr. Fink! – designs most often call for a Designer] Likewise, the sleeping eternal universe 'engineered' us in its kerzillion years spinning journey."

What does Mr. Fink intend by this last sentence? Does he mean the eternal universe itself is sleeping as though it had no brain to manufacture us? And that is, indeed, the issue. Fink's "thinking" analogy (the universe in this case) continues to break down as it did with the seeds. Seeds and the universe have no thinking power and therefore are not analogous to his argument.

The word *engineered* is in quotation marks to alert the readers: it was impossible for the universe to manufacture us since the origin of life was not an intended design by any Creator but came as the result of blind chance. Fink's contention is: The universe did not "engineer" us since this, in turn, would have called for an *Engineer* who masterminded and constructed the *glory* of the universe. Fink believes in the naturalist's formula for the origin of the universe: *Nobody times nothing equals everything* (MacArthur).

Fink continues to write: "But the infinity of an always-existing creator (and with no universe to live in until it was created by the creator) can be imagined by creationists, then why can't the plainer infinity be imagined; of an always existing and swirling dust, rubble and churning fires which makes up our lawful, yet blind universe?"

These are Fink's final remarks but he waits until the end of his essay to make any sense whatsoever. I feel that Fink makes an excellent point and it should be well taken and considered by all creationists: It is easier to imagine the plainer infinity of our universe "always existing" from "swirling dust, rubble, and burning fires" than it is to imagine "the infinity of an always-existing creator."

Before speaking to the issue of God and the universe or – in the thinking of Fink – no God and the universe, let us get back to "dumb" Charlotte. At this point, let us see where we are with understanding Fink's way of thought!

Firstly, the making of Charlotte's web was not due to the *intelligence of the spider*. In fact, Fink has the following nomenclature for his favorite spider:

- "Poor retard"
- "comatose idiot"
- "dumb Charlotte"

Secondly, the making of Charlotte's web was not due to the *intelligence of a Creator* because there is no "intelligence behind it." What is more, Fink implies the so-called marvels of life need no Creator. Then how does Fink account for spider webs that "are complex chemical engineering marvels"? Simply by loading us with some *heavy science*, informing us that Charlotte has "genetically programmed actions ..."

I must admit: I do not know much about the thinking power of animals or the various levels of intelligence manifested from the lowest of creatures up to mankind but I cannot believe that all nature is filled with "comatose idiots" and "poor retards" who are automated to carry out "actions" in such a mechanized manner that there is no room for, at the very least, a simple reflection.

I perceive Mr. Fink, since he has no faith in a Creator, believes in evolution. Therefore, he must account for the following facts that he mentions in his essay:

- "Spider webs are complex chemical engineering marvels. Why? Because spiders use the least materials, the fewest stands necessary and cross the shortest distances to produce a nearly weightless, nearly invisible insect trap – the largest and strongest know structure considering the spider's size."

- "It builds [the web] unaided, and ... it must carry and produce all its own building materials.

- "... Because daily, each web was differently adapted to changed circumstances."

- "... The re-spun web had a very different configuration ..."

Stop me, if I am wrong! In presenting the above facts, Fink is arguing from the wrong side. He reminds me of a creationist, presenting facts from the standpoint of intelligent design and a Creator; not an evolutionist from the perspective of no intelligent design and no God. He ends his observation with, "so, there's an incredibly 'intelligent-looking' marvel: The glorious web, with fabulously integrated efficiency and adroit re-adaptation to surrounding changes, yet all produced by a comatose idiot!" All the marvels preceding Fink's favorite expression from the annals of artistic literary expression – "comatose idiot" – are self-defeating to his position. The patterns of a spider and her web-building; such observations overcome his entire argument.

What Fink calls "actions" of the spider are termed "behavior patterns" by scientists whose expertise is in the realm of nature. How can Fink explain the origin of these patterns? His spider repertoire involves several complex steps, each one being crucial to the whole ritual of web building. These complex behavior patterns defy plausible evolutionary explanations. The most amazing fact: should Charlotte have spiderlings, they too will perform identical actions or behavior patterns as their mother. Science has proved that learned knowledge is not biologically passed from one generation to the next. Then what is the solution to the issue of behavior patterns?

The way spiders know how to live and survive is called instinct. Contrary to what Fink writes, instinct *is knowledge or know-how* which is programmed into a creature at birth. Man has never been able to explain how this development could have come about through the process of evolution. Fink comments that Charlotte is not aware "of her genetically programmed action ..." That is a correct assumption but such programmed knowledge demands an Intelligent Programmer and, therefore, testifies to a Creator. Charlotte is, of course, not aware she has been programmed genetically but carries out her purposes just the same.

Being the arm-chair scientist that Fink undoubtedly thinks he is, when he writes of Charlotte's "genetically programmed action" I would suppose that biochemical genetics is a part of his scientific deftness. However, in his concluding these entire spider marvels are the result of genitival programming, I can not help but feel that Fink is attributing some process of magic to this type of programming without being fully aware of its implications.

Mr. Fink! What do you think, in your estimation, to be the origin of the genetic code or DNA? I don't suppose you have the slightest idea but you are not alone in this quandary.

The following statements are the words of scientists who specialize in molecular biology. Leslie E. Orgel is one of the top biochemists in the world and of special repute in origin-of-life studies. He writes:

"We do not yet understand even the general features of the origin of the genetic code ... the origin of the genetic code is the most baffling aspect of the problem of the origins of life, and a major conceptual or experimental breakthrough may be needed before we can make any substantial progress."

"Darwinism at the Very Beginning of Life" *New Scientist,* vol.94 (April 15, 1982) pp. 149-152.

One more quote of many that could be sighted. Haskins comments:

"But the most sweeping evolutionary questions at the level of biochemical genetics are still unanswered. How the genetic code first appeared and then evolved and, earlier than that, how life itself originated on earth remain for the future to resolve ... The fact that in all organisms living today the processes both of replication of the DNA and of the effective translation of its code requires highly precise enzymes and that, at the same time the molecular structures of those same enzymes are precisely specified by the DNA itself, poses a remarkable evolutionary mystery .... Did the code and the means of translating it appear simultaneously in evolution? It seems almost incredible that any such coincidence could have occurred, given the extraordinary complexities of both sides and the requirement that they be coordinated accurately for survival. *By a pre-Darwinian (or a skeptic of evolution after Darwin) this puzzle would surely have been interpreted as the most powerful sort of evidence for special creation.*" [Italics, mine]

Caryl P. Haskins, "Advances and Challenges in Science in 1970," *American Scientist*, vol.59 (May/June 1971), p.298-307.

Mr. Fink! Continuing with further appeal to your reasoning powers; consider some of the body parts of charlotte! :

She has in her cephalothoraxes: eyes, mouthparts and four pairs of legs which are protected by a shield like covering, the carapace. Her abdomen contains the respiratory, digestive, and reproductive organs as well as the spinnerets, which spin silk with just the right chemical consistency. Charlotte, as she grows, sheds or molts her exoskeleton several times. To do this, in one of her factories, she must produce a molting liquid that dissolves her shell. The stages and molts that she goes through is a "simple metamorphosis" but she must do this in order to live.

What do you think are the odds, in evolution's game of chance, of the behavior patterns of a spider coming about through natural selection? Evolution can not account for even one cell in Charlotte's body let alone her factories for producing silk and molting fluid.

James F. Coppedge, a research analyst in molecular biology, writes this appeal:

"Let us consider this question: By all the rules of reason, could there be a code which carries a message without someone originating that code? It would seem self-evident that any such complex message system, which is seen to be wise and effective, requires not only intelligence but a person back of it. Who wrote the DNA code? Who is the author of this precise language? There is no evolutionary explanation that even begins to be an adequate answer. The only logical thing to do is to listen to the voice of reason and to acknowledge that only God, the infinite Person, could author that amazing *living language*."

Evolution: Possible or Impossible?

P.138

Mr. Fink! You are chauvinistic against creationists and you allege they have *only faith* to "prove" their creation theories, claiming creationists follow "a make-it-up-as-you-go 'science'" but the fact is: coupled with their faith, there is a plethora of empirical evidence to back them up. Perhaps your view of DNA is a make-it-up-as-you-go-along "science" since you have neither bothered to explain its origin nor how such a living language can possibly exist without God.

We are now ready to move to Appendix 3.

Appendix III
THE STEADY-STATE THEORY

"In the beginning God created the heavens and the earth."

Genesis 1:1

NIV

"Do you not know? Have you not heard? The LORD is the everlasting God, the creator of the ends of the earth."

Isaiah 40:28

NIV

Mr. Fink! Again, I would appeal to your reasoning power regarding your *no-God universe*. I am rather surprised at your words of "an always existing ... yet blind universe." Apparently, you believe in the old theory of an eternal universe called the Steady State theory. The more recent cosmic theory of the Big Bang, I thought, would be more in your ball-park.

At least there is some evidence for the support of the Big Bang. But finding valid, substantial support for either cosmic model (The Big Bang or the Steady State) of the universe has too many contradictions. However, for the purpose at hand, I will stick with your belief and will deal with the theory to the best of my ability.

In case my readers are not familiar with the Steady State theory, the following is a brief description:

The universe has always existed and will exist for ever; so that as old galaxies die, material is continuously created out of nothingness, in the form of hydrogen atoms. However, most astronomers, physicists, and cosmic philosophers agree that this idea in no way provides a solution to the problem of the actual creation of the universe.

I will begin by noting that your essay, in stating your belief of the universe is quite logical and well thought out:

"But if the infinity of an always-existing creator (and with no universe to live in until it was created by the creator) can be imagined by creationists, then why can't the plainer infinity be imagined: of an always existing and swirling dust, rubble and churning fires which make up our lawful, yet blind universe?"

You make an excellent point: If creationists are able to imagine an ever-existing God, then they should be able to imagine an ever-existing universe. I would assume by "plainer" infinity, you intend "Isn't it more scientific to believe in a natural-caused and materialistic universe rather than a theological one which includes a Creator-God?"

In 1998, in the CEN *Technical Journal*, 12 (1): 20-22, J. Sarfati wrote "If God created the universe, then who created God?" You and Sarfati share the same question: "'Who' created the Creator?" You even have the pronoun *who* in quotes, to indicate that it takes an identity to create God who, after all, couldn't possibly have sprung out of nothingness. Your thought is cynical and yet *you are able to acknowledge cause and effect*. More information will be given about this as we get involved with our subject.

Above, I mentioned two theories: The Steady State and the Big Bang. The Steady State cosmic theory is out-dated by the Big Bang cosmic theory and for the following reason:

Astronomers, through their discovery of back-ground radiation, the spectra of nebulae "shifting" toward the red end of the electromagnetic spectrum, have concluded that the "proof" of the expanding universe sounded the death knell for the Steady-State or Ever-lasting Universe. In other words, most astronomers are convinced that the universe is expanding and that this implied a beginning. A few are still holding out for an *eternal past* and so, you are not alone.

With all this having been said, I don't disagree with your belief on the basis of the Big Bang since I feel that would be overcoming a bad theory with a theory equally as bad, if not worse. I disagree with your theory not because the proponents of the Big Bang claim the universe had a beginning but because of their doctrine of the constant renewal of hydrogen atoms created out of material from nothingness. I view this as contradictory to the fundamental laws of the physical sciences; the laws of thermodynamics.

These laws indicate to me the universe, in deed, had a beginning but the universe is dying from heat loss:

- The FIRST LAW states: the total amount of mass-energy in the universe is *constant*.

- The SECOND LAW states: the amount of energy in the universe available for work is running down, or *entropy is increasing to a maximum.*

These empirical laws of science state: the total amount of mass-energy is limited and the amount of usable energy is decreasing. In other words, not only is the universe dying from heat loss but according to the second law of thermodynamics (Entropy): everything runs relentlessly from order to disorder and from complexity to decay.

These facts are crucial in my decision for believing the universe had a beginning and in opposition to the belief; the universe is eternal and renews itself by forming hydrogen atoms, etc. If the total amount of mass-energy is limited and the amount of usable energy is decreasing, then the universe could not have existed forever otherwise it would already have exhausted all usable energy and reached what is known as heat death.

[Special Side Note for Creationists: There are a few Creationists who believe the Second Law of Thermodynamics was not in operation before the fall of man. In a universe were the sources of accessible energy can run down, I hardly believe this Law has anything to do with Adam's sin. It was part of God's natural laws. I assume that Adam's need to partake of food from the tree of life was not only for pleasure but to meet the needs of a physical body that requires a constant source of energy and vitality. Would not this imply that our first parents were under the Second Law of Thermodynamics before their fall into sin? To believe otherwise is to accept Adam and Eve's *innate immortality*, which is definitely *not a biblical doctrine.* Immortality was granted on the condition of obedience to the moral law (Romans 6:23; 5:12; I John 3:4).After sin, holy angels were commissioned to guard the tree of life. Hence, there are no immortal sinners. Following transgression, our first parents were deprived of the tree of life; had their vitality diminished and gradually died. The same creationists who claim the Second Law of Thermodynamics came into operation after the fall of Adam do, in fact, teach that Adam had INNATE IMMORTALITY. This doctrine is the natural consequence of believing the first part of the previous sentence. These creationists reason that because man was created in the image of God, he therefore was IMMORTAL. They set up the following syllogism:

(Major) First Premise: God is IMMORTAL – I Timothy 1:17.

(Minor) Second Premise: Man was created in God's Image – Genesis 1:27.

Conclusion: Therefore, man is IMMORTAL.

Such reasoning can be *proved to be unsound and fallacious* by setting up a second syllogism:

(Major) First Premise: GOD is ALL-WISE, ALL POWERFUL, and WHOSE PRESENCE CAN BE EVERYWHERE.

(Minor) Second Premise: Man was created in God's Image.

Conclusion: MAN is ALL-WISE, ALL POWERFUL, and WHOSE PRESENCE CAN BE EVERYWHERE.

The above characteristics of God are some of His distinctive attributes. IMMORTALITY IS THE ONLY ASPECT OF HIS POWER GRANTED TO MAN AND IT IS NOT INNATE IN MAN NOR IS IT AWARDED UNCONDITIONALLY TO MAN. It comes through faith and by God's grace. That is why Paul has written:

"Fight the good fight of faith. Take hold of eternal life … In the sight of God … *Who alone is immortal* and who lives in unapproachable light, which no one has seen or can see."

I Timothy 6: 12, 13, 16

NIV

[Eternal life is a part of the resurrection promise granted to man through redemption – I Corinthians 15: 45-56]

Before leaving this block of thought, the same creationists who teach the immortality of the soul also teach that Genesis 2:7 is the formula for instructing us that God placed an immortal soul within man's body but no matter how I read this scripture and in whatever version I read it, there is no mathematical equation that ever equals IMMORTAL SOUL. Let us read the scripture:

" …. The LORD God formed man from the dust of the ground and breathed into his nostrils the breath of life, and the man became a living being."

Genesis 2:7

NIV

Some creationists see the following formula:

Dust of the ground + the breath of life = An IMMORTAL SOUL.

But this is reading into the scripture something that is not there. Exegesis is to give light on the text – not to obscure its meaning by substituting "immortal soul" into the context. The following is the way the formula should be read and no Bible Commentaries are needed to make clear what the Bible has already made obvious:

Dust of the ground + the breath of life = a living being.

No matter what word substitutes are used, the formula works out to the same conclusion:

Body + Breath = a person.

The chemical constituents of the ground formed into a human body + the spark of life from God = the living entity called man.

James puts it this way: "As the body without the spirit is dead, so faith without deeds is dead."

James 2:26

NIV

The "spirit" [*pneuma*] in the Greek New Testament is the same as "breath" [*ruach*] in the Old Testament. In the Septuagint, *ruach* is always rendered *pneuma*. Therefore:

A body + breath = a LIVING body.

A body – breath = a DEAD body, just as James wrote.

OR:

A body with the spirit is LIVING.

A body without the spirit is DEAD, just as James wrote.

Again, Eternal Life is part of the Resurrection Promise granted to man through redemption – I Corinthians 15: 45–56; II Timothy 4: 6–8]

Mr. Fink! It was alright for you to read my Special Side Note for Creationists but at this point, your essay comes into play – "'Who' created the Creator?" You placed quotes around the pronoun *who*. You seem to be cognizant of cause and effect even though your question is sarcastic. I have already indicated why the universe had a beginning but allow me to set up the idea in another syllogism:

(Major) First Premise: Everything which has a beginning has a cause.

(Minor) Second Premise: The universe has a beginning.

Conclusion: Therefore, the universe has a cause.

You made the statement: "swirling dust, rubble, and churning fires" make up our always existing, lawful, yet blind universe. But it is not so blind to its own laws that it would disallow men to probe its mysteries by the discovery of thermodynamics. I understand the EFFECTS of your universe; the swirling dust, rubble, and churning fires but what is the CAUSE? Where did these forces of nature come from? What is their source?

David Darling, "On Creating Something from Nothing," *New Scientist*, vol.151 (September 14, 1996) writes:

"What is a big deal – the biggest deal of all – is how you get something out of nothing.

"Don't let cosmologists try to kid you on this one. They have not got a clue either – despite the fact that they are doing a pretty good job of convincing themselves and others that this is really not a problem. 'In the beginning,' they will say, 'there was nothing – no time, space, matter or energy. Then there a was a quantum fluctuation from which …' Whoa! Stop right there. You see what I mean? First there is nothing, [and] then there is something. And the cosmologists try to bridge the two with a quantum flutter, a tremor of uncertainty that sparks it all off. Then they are away and before you know it, they have pulled a hundred billion galaxies out of their quantum hats." ….

P.49

"You cannot fudge this appealing to quantum mechanics. Either there is nothing to begin with, in which case there is no quantum vacuum, no pregeometric dust, no time in which anything can happen, no physical laws that can effect a change from nothingness into somethingness; or there is something, in which case that needs explaining."

Ibid. P.49

In contemplating the above remarks; they only make me more convinced and makes it easier for me to imagine an Eternal Creator (Isaiah 9:6 and 40:25-28; Colossians 1:16-17; Hebrews 1:1-3) who, by creation, was the effectual CAUSE for this universe and everything we see within it, than to believe that the stars, galaxies, and solar system arose without cause and from nothingness.

You should understand by David Darlings statement: a quantum fluctuation is not the answer to the problem on how to get something from nothing.

I have one more syllogism for you to think about. You know about radioactivity and its decay stages. The facts of radioactivity very positively forbid the past eternity of matter for again, in that radioactivity is an indication of entropy, it must have died out an "eternity ago."

In that scientists know of no "natural" means for producing mass I am enabled to set up the following syllogism:

(Major) First Premise: Matter is not being brought in existence by *present* "natural" means.

(Minor) Second Premise: The facts of radioactivity negate the *past eternity* of matter.

Conclusion: Therefore, matter was brought into existence at some time *in the past* by means equivalent to *a real Creation*.

[This affirms GENESIS 1:1 – "In the beginning God created the heavens and the earth." KJV]

Mr. Fink! I'm unable to prove God exists; you can't prove that He doesn't exist. But who is left in a *greater quandary*? Which reasoning stands up to *greater scrutiny*? Let me compare our beliefs concerning the universe:

[RICHARD: The universe can be shown to have had a beginning (This is made plain by the laws of thermodynamics).

It is unreasonable to believe something could begin to exist without a cause. Therefore, I believe in a CAUSE of the things that I witness in the universe and the natural world – that is, the EFFECTS which sprang from an event other than nothingness).

The universe requires a CAUSE and I believe that CAUSE is not something (as quantum fluctuation) but SOMEONE – the personal and eternal God of Genesis 1:1 – "In the beginning God created the heavens and the earth."]

[BOB: The universe is from eternity. It has always existed and will exist for ever (The empirical laws of Thermodynamics can be disregarded).

It is reasonable to believe something could begin to exist without a cause. This is why I believe in the Steady State theory: the universe is from eternity and swirling dust, rubble and churning fires came into being without cause or purpose. Matter took its form from nothingness.

The universe doesn't require a cause because it has no God and it exists by chance, uncertainty, and from nothingness.]

The question we need to ask ourselves is this: What belief is best suitable for man's happiness and harmonious existence on this planet? – A materialistic and atheistic belief or a spiritual and theological belief?

Mr. Fink! I would invite you to read the Bible for direction in your thinking on these matters. The Bible actually taught the *scientific law of entropy* thousand of years before empirical science systematized this conception of the universe into a natural law. The prophet Isaiah and king David both declared that the heavens and earth would "wear out like a garment." (ISAIAH 51:6; PSALMS 102:25-26 KJV). In the first century, the apostle Paul looked forward to the day when "the creation itself will be liberated from its bondage to decay" (ROMANS 8:21 KJV).

There is one other item in your essay that deserves attention. You wrote "But if the infinity of an always-existing creator (and with no universe to live in until it was created by the creator) can be imagined by creationists ...."

I spent some time thinking about your observation. The majority of creationists do believe (me included) that our present universe is only thousands of years old. What, then, was God doing prior to *this creation*?

Bob! This is why I believe that God created other universes and planets with their inhabitants. I don't believe in little green men with enlarged eyes or those creatures conjured up by science-fiction writers who write about Mars and other planets.

However, I do believe in unfallen worlds with people living under life-sustaining conditions similar to ours. (See HEBREWS 1:1, 2; REVELATION 12:10-12; ISAIAH 40:22: ROMANS 8:22, 23; NEHEMIAH 9:6)

A religious writer has written of our world:

"He (Christ) has left the courts of heaven, where all is purity, happiness, and glory, to save the one lost sheep, the one world that has fallen by transgression. And He will not turn from His mission. He will become the propitiation of a race that has willed to sin."

The Desire of Ages [Luke 15: 1-7; John 10: 1-18]

Ellen G. White

P.693

Final thoughts in this Appendix will be directed to Albert Einstein. In the first stages of his theory, Einstein made general relativity analogous to the Static State theory – the universe had always existed. However, it was later in time that he became aware, his own theory predicted that the universe had a beginning. Einstein, in recognizing this information, aspired "to know how God created the world." The bottom line was: Einstein concluded that the CAUSE of the universe was GOD.

Above, I pointed out that thermodynamics and especially its second law of entropy, proves that the universe had a beginning. Einstein said:

"Classical thermodynamics ... is the only physical theory of universal content concerning which I am convinced that, within the framework of applicability of its basic concepts, it will never be overthrown." ["Thermodynamics in Einstein's Thought" *Science*, vol.157, August 4, 1967, Martin J. Klein, p.509]

In referring to the Entropy Law, Einstein said; "it is the premier law of all science." Einstein came to support two crucial things about the universe:

- GOD was the CAUSE of the UNIVERSE.

- The UNIVERSE *did not eternally exist; it had a beginning.*

Mr. Fink! You have an opportunity to cast your lot on the side of Albert Einstein who was, no doubt, the greatest physicist in the world prior to his death in 1955.

But long before Einstein, the apostle Paul (an intellectual marvel of the first century) wrote:

"For since the creation of the world God's invisible qualities – his eternal power and divine nature – have been clearly seen, being understood from what has been made, so that men are without excuse" (Romans 1:20 KJV).

Paul makes an appeal to your rational judgment by claiming that the universe and our solar system could not have come from nothing. He writes that men see the *effects* of some event and since there is no natural explanation for mass and energy, the supernatural cause of this event that Paul named creation, necessarily calls for a Creator – *God himself.*

In this appendix, I have responded to issues that were only applicable to the words in your essay. In so doing, I have been one-sided with the law of entropy and how it speaks of decay and death. I have not mentioned the optimistic side of my belief concerning God's eternal kingdom.

But a hint of this optimism and God's eternal purposes are intimated in the prophecy of Isaiah:

"For, behold, I [God] create new heavens and a new earth: and the former shall not be remembered, nor come into mind" (Isaiah 65:17 KJV).

King David said, "The heaven, even the heavens, are the Lord's: but the earth hath he given to the children of men" (Psalm 115:16 KJV).

The exhortation of those prophetic words is echoed in The Sermon on the Mount:

"Blessed are the meek: for they shall inherit the earth" (Matthew 5:5 KJV).

The apostle Paul, in his words concerning entropy, wrote:

"Because the creature [creation] itself also shall be delivered from the bondage of decay in the glorious liberty of the children of God" (Romans 8:21 KJV).

The apostle Peter gave the warning and admonition:

"But the day of the Lord will come as a thief in the night; in which the heavens shall pass away with a great noise, and the elements shall melt with fervent heat, the earth also and the works that are therein shall be burned up …. ["The Lord will come as a Thief in the night" not because His second coming will be secretive (Matthew 24: 30-31) but because sinners, who are totally involved in this world's affairs, will be completely unprepared for the Lord's coming]

"Nevertheless we, according to his promise, look for new heavens and a new earth, wherein dwelleth righteousness" (2 Peter 3: 10, 13 KJV).

Finally, John writes:

"And I saw a new heaven and a new earth: for the first heaven and the first earth were passed away; and there was no more sea. And I John saw the holy city, New Jerusalem, coming down from God out of heaven, prepared as a bride adorned for her husband. And I heard a great voice out of heaven saying, Behold, the tabernacle of God is with men, and he will dwell with them, and they shall be his people, and God himself shall be with them, and be their God. And God shall wipe away all tears from their eyes; and there shall be no more death, neither sorrow, nor crying, neither shall there be any more pain: for the former things are passed away" (Revelation 21: 1-4 KJV).

Mr. Fink! Most of the elements in the earth have been exploited by man or have been used for the good purposes of man and his survival. I do not know about God's plan for the new earth, about the re-making of elements or the re-establishing of the laws of thermodynamics. But I do know that whatever is in the plan of God, this earth, forever, will be our home: "God himself that formed the earth and made it; he hath established it, he created it not in vain, he formed it to be inhabited" (Isaiah 45:18 KJV). In the last part of this verse God claims, "I am the LORD; and there is *none else.*" [Indicating that *God was not created*]

Sincerely, Richard B. Pittack

SOURCES
BOOKS (Single Author)
Creationism:

Coffin, Harold, Creation – *Accident or Design?* Review and Herald Publishing Association, Washington, D.C., 1969. Copyright © 1969 by the Review and Herald publishing Association.

Coppedge, James F., *Evolution: Possible or Impossible?* Santa Clarita, California published by Probability Research in Molecular Biology.

Comninellis, Nicholas, *Creative Defense*, First Printing: September 2001, Green Forest, AR, Copyright © 2001 by Master Books.

Denton, Michael, *Evolution: A Theory in Crisis*, Adler & Adler, Publishers, Inc., Chevy Chase. MD, 1985, Copyright © 1985 by Michael Denton. [A book written by an evolutionist but it has very few thoughts in disagreement with creationists]

Gish, Duane T., *Creation Scientists Answer Their Critics*, Institute for Creation Research, El Cajon, CA, First Edition, 1993, Copyright © 1993.

Gish, Duane T., *Evolution*: *The Fossils Still Say No!* Institute for Creation Research, El Cajon, California, First Edition Revision, Copyright © 1995.

Hanegraaff, Hank, *The Farce of Evolution*, World Publishing, Nashville, TN, 1998, Copyright © 1998 by Hank Hanegraaff.

Hayward, Alan, *Creation and Evolution,* 1985 Copyright © Alan Hayward, Minneapolis, Minnesota, Bethany House Publishers.

Klotz, John W., *Studies in Creation*, 1985, St. Louis, MO, Copyright © 1985 Concordia Publishing House.

Marsh, Frank Lewis, *Life, Man, and Time*, Outdoor Pictures, Escondido, California, Revised edition, 1967, Copyright, 1967, by Frank Lewis Marsh.

Morris, Henry M., *Science and the Bible*, Revised and Updated, © 1986 by Henry M. Morris, Moody Press Chicago.

Morris, Henry M., *That Their Word May Be Used Against Them* (Quotes from Evolutionists Useful for Creationists) Compiled by Henry M. Morris, 1997, San Diego, CA, Institute for Creation Research.

Perloff, James, *Tornado in a Junkyard*, Refuge Books, Arlington, MA, third printing, July 2000, Copyright © by James Perloff.

Ritland, Richard M., *A Search for Meaning in Nature,* Pacific Press Publishing Association, Mountain View, California, 1970, Copyright © 1970, by Pacific Press Publishing Association.

Sunderland, Luther D., *Darwin's Enigma*, Master Book Publishers, Santee, California, Second Edition, 1984, Copyright © 1984, Luther D. Sunderland.

Wells, Jonathan, *The Politically Incorrect Guide to Darwinism and Intelligent Design*, Regnery Publishing Inc. An Eagle Publishing Company, Washington, DC, Copyright © 2006 by Jonathan Wells.

BOOKS (More Than One Author)
Creationism:

Ankerberg, John & Weldon, John, *Darwin's Leap of Faith*, Harvest House Publishers, Eugene, Oregon, 1998, Copyright © 1998 by John Ankerberg and John Weldon.

Morris, Henry M., Morris, John D., *Modern Creation Trilogy*, (vol.2) Science and Creation, Master Books, second printing, November, 1997, Copyright © 1996 by Master Books.

BOOKS (Single Author)
Evolutionism:

Darwin, Charles, *The Origin of Species*, (1859) and *The Decent of Man* (1871). Bennett A. Cerf, Donald S. Klopfer, The Modern Library, Random House, Inc., New York.

Gould, Stephen Jay, *Ever Since Darwin*, Copyright © 1977 by Stephen Gould, W. W. Norton & Company, New York and London.

Gould, Stephen Jay, *The Panda's Thumb*, W. W. Norton & Company, Inc., New York, N.Y., 1980, Copyright © 1980 by Stephen Jay Gould.

Gould, Stephen Jay, *An Urchin in the Storm*, W. W. Norton & Company, New York and London, Copyright © 1987 by Stephen Jay Gould.

Gould, Stephen Jay, *Bully for Brontosaurus*, W. W. Norton & Company, Inc., New York and London, N.Y., 1991. Copyright © 1991 by Stephen Jay Gould.

Gould, Stephen Jay, *Wonderful Life*, W. W. Norton and Company, New York and London, Copyright © 1989 by Stephen Gould.

Gould, Stephen Jay, *Dinosaur in a Haystack*, Harmony Books, a division of Crown Publishers, Inc., New York, New York. Copyright © 1995 by Stephen Jay Gould.

Gould, Stephen Jay, *Leonardo's Mountains of Clams and The Diet of Worms*, Three Rivers Press is a registered trade mark of Random House, Inc., New York, New York. Copyright © 1998 by Turbo, Inc.

Gould, Stephen Jay, *I Have Landed*, Harmony Books, New York, New York, Copyright © 2002 by Turbo, Inc.

Isaak, Mark, *The Counter-Creationism Handbook*, University of California Press (Berkeley. Los Angeles. London) © 2005, 2007 by Mark Isaak.

Lanham, Url, *The Bone Hunters*, Dover Publications, Inc., New York, Copyright © 1973 by Url Lanham.

Peterson, Roger Tony, *World Atlas of Birds*, Crown Publishers Inc., New York, Copyright © Roger Tony Peterson.

BOOKS (More Than One Author)
Evolutionism:

Gribbin, John and Cherbas, Jeremy, *The First Chimpanzee in Search of Human Origins*, Barnes & Noble Books, New York, Copyright © 2001 by John Gribbin and Jeremy Cherfas.

Rayner, J.M.V., *Biomechanics in Evolution*, Ed J.M.V. Rayner and R.J. Wooten (Cambridge; Cambridge University Press, 1991, P.194.

BOOKS (Single Author)
Dinosaurology
Evolutionism:

Bakker, Robert T., *The Dinosaur Heresies*, Zebra are published by Kensington Publishing Corp., Copyright © 1986 by Robert T. Bakker.

Charig, Alan, *A New Look At Dinosaurs*, Facts On File, Inc., Copyright © 1979, 1983 British Museum (Natural History) Reprinted 1985, 1988.

Colbert, Edwin H., *The Great Dinosaur Hunters and Their Discoveries*, Dover Publications, Inc., New York, Copyright © 1968, 1984 by Edwin H. Colbert.

Desmond, Adrian J., *The Hot – Blooded Dinosaurs A Revolution In Paleontology*, The Dial Press/James Wade, New York, 1976, First published in Great Britain by Blond & Briggs Ltd., Copyright © 1975 by Adrian J. Desmond.

Fiffer, Steve, *Tyrannosaurus Sue*, W. H. Freeman and Company, New York, © 2000 by Steve Fiffer.

Gillette, David D., *Seismosaurus The Earth Shaker*, with illustrations by Mark Hallett, Columbia University Press, New York, Copyright © 1994 Columbia University Press, © 1993 Mark Hallett.

Jacobs, Louis, *Quest for the African Dinosaurs*, Villard Books, New York, 1993, Copyright © 1993 by Louis L. Jacobs.

Jaffe, Mark, *The Gilded Dinosaur*, Three rivers Press, New York, Copyright © 2000 by Mark Jaffe.

Krishtalka, Leonard, *Dinosaur Plots and Other Intrigues In Natural History*, Avon Books, New York, Copyright © 1989 by Leonard Krishtalka.

Lessem, Don, *Time for Learning Dinosaurs*, Peter Dodson Consultant, Illustrator, Phil Wilson, Copyright © 2004 Publications International, Ltd.

Lessem, Don, *Kings of Creation*, Illustrated by Jon Sibbick, Simon & Schuster, New York, New York, Copyright © 1992 by Don Lessem.

McGowan, Christopher, *Dinosaurs, Spitfires, & Sea Dragons*, Harvard University Press, Cambridge, Massachusetts, London, England, 1991, Copyright © 1983, 1991 by Christopher McGowan.

Norman, David, *The Prehistoric World of the Dinosaur*, Gallery Books, an imprint of W. H. Smith Publishers Inc., New York, New York, Copyright © 1988 Brompton Books Corp

Norman, David, *The Illustrated Encyclopedia of Dinosaurs*, Published by Crescent Books, New York, © Salamander Books Ltd 1985.

Norman, David, *Dinosaur!* , Prentice Hall General Reference, New York, New York, Text Copyright © 1991 by David Norman.

Novacek, Michael, *Dinosaurs of the Flaming Cliffs*, Anchor Books Published by Doubleday, New York, New York, Copyright © 1996 by Michael Novacek.

Paul, Gregory S., *Predatory Dinosaurs of the World*, (A New York Academy of Sciences Book written and drawn by Gregory S. Paul), A Touchstone Book published by Simon & Schuster Inc., New York, Copyright © 1988 by Gregory S. Paul.

Spalding, David A. E., *Dinosaur Hunters*, Prima Publishing, Copyright © 1993 Brandywine Enterprises B.C. Limited.

BOOKS (More Than One Author)
Dinosaurology
Evolutionism:

Svarney, Thomas E. and Svarney-Barnes, Patricia, *The Handy Dinosaur Answer Book*, Visible Ink Press, Farmington Hills, MI, Copyright © 2000 by visible Ink Press ®

The Nature Companions Rocks, Fossils and Dinosaurs, Consultant Editors – David Roots and Paul Willis (ROCKS AND FOSSILS), Michael K. Brett-Surman (DINOSAURS), Published by Fog City Press, San Francisco, CA, Copyright © 2002 Weldon Owen Pty Ltd. [Quote taken from the second chapter by Christopher A. Brochu and Colin McHenry entitled, "The World in the Age of Dinosaurs", P.296.

Lambert, David and the Diagram Group, *The Dinosaur Data Book*, Avon Books, New York, New York, Copyright © 1990 by David Lambert and Diagram Visual Information Ltd.

Lambert, David and the Diagram Group, *A Field Guide to Dinosaurs*, Avon Books, New York, New York, Copyright © 1983 by Diagram Visual Information Ltd.

Czerkas, Sylvia J. and Czerkas, Stephen A., *Dinosaurs a Global View*, First published by Dragon's World 1990, © Dragon's World Ltd 1990 © test Sylvia J. and Stephen A. Czerkas 1990 © Colour Artwork resided with the Individual artists 1990.

Gardom, Tim with Milner Angela Scientific Advisor, Prima Publishing, Rocklin, CA, Copyright © 1993 *The Book of Dinosaurs The Natural History Museum Guide*, London.

Psihoyos, Louie with Knoebber, John, *Hunting Dinosaurs*, Random House, Inc., New York, Copyright © 1994 by Louie Psihoyos.

BOOKS (Geology)
Creationism:

Clark, Harold W., *Fossils, Flood, and Fire*, OUTDOOR PICTURES, Escondido, California, Copyright, 1968, Harold W. Clark.

Coffin, Harold G., *Earth's Story*, Washington, D.C., Copyright © 1977, Review and Herald Association.

Price, George McCready, *Common-Sense Geology*, Copyright 1946 by the Pacific Press Publishing Association.

Price, George McCready, *The Story of the Fossils*, Copyright, 1954, by Pacific Press Publishing Association.

Wheeler, Gerald, *Deluge*, Copyright © 1978 by Southern Publishing Association.

Morris, Henry M., Whitcomb, John C., *The Genesis Flood*, Philadelphia, Pennsylvania, Copyright, 1961, by the Presbyterian and Reformed Publishing Company.

Genesis and Geology, Two Chapters on Science Reprinted from the SEVENTH-DAY ADVENTIST BIBLE COMMENTARY volume 1, Washington D.C., Copyright 1978 by Review and Herald Publishing Association.

Numbers, Ronald L., Editor, *Selected Work of George McCready Price*, Introductions copyright © 1995 Ronald Numbers, New York & London, Garland Publishing, Inc.

Woodmorappe, John, *Studies in Flood Geology*, Second Edition, El Cajon, California, © 1999 by The Institute for Creation Research.

BOOKS (Geology)
Evolutionism:

Longwell, Chester R.; Knopf, Adolph; Flint, Richard F*., Physical Geology*, Third Edition, Copyright, 1948 by New York, John Wiley & Sons, Inc.

Shimer, John A., *This Sculptured Earth: The Landscape of America*, Copyright © 1959 by, New York, Columbia University Press.

Bates, Robert L.; Jackson, Julia A., editors, *Dictionary of Geological Terms*, Prepared under the direction of the American Geological Institute, Copyright © 1984 by the American Geological Institute.

NEWSPAPERS:

British Newspapers – *The Guardian Weekly*, 26 November 1978, vol. 119, no 22 Pl.

Martin, Larry D., "The Barosaurus Is No Five-Story-Tall Canary", *Sunday World – Herald*, Omaha, Nebraska, 19 January 1992, p. B-17.

FAQs:

Center for Scientific Creation (1995-1997) issued FAQs entitled "What Was Archaeopteryx "– a five paged document.

http://www.creationscience.com/onlinebookfaq/archaeopteryx.shtml

"Charlotte's Web" (December 11, 2004)
http://www.webster.sk.ca/greenwhich/bible-a.htm

ARTICLES:

Benton, M. J., (*Archaeopteryx* was single-headed, as in reptiles) *Nature*: 305:99 (1983).

Behrensmeyer, Anna K., "Taphonomy and the Fossil Record," *American Scientist*, vol. 72 (November, December 1984), p.560.

Charig, Alan J., Frank Greenaway, Angela C. Milner, Cyril A. Walker, and Peter J. Whybrow, "*Archaeopteryx* Is Not a Forgery," *Science*, vol.232 (May 2,1968), pp. 622-626.

"A Review of Claims about *Archaeopteryx* in the Light of the Evidence," *Creation Research Society* Quarterly, June 1995, p.18

Dodson, Peter, "International Archaeopteryx Conference," *Journal of Vertebrate Paleontology*, vol.5 (June 1985), pp. 177-179.

Gibbons, Ann, "New Feathered Fossil Brings Dinosaurs and Birds Closer," *Science*, vol. 274 (November I, 1996), pp. 720-721.

"Search for the Holy Transformation," review of Evolution of Living Organisms, by Pierre-P. Grasse, *Paleobiology*, vol.5 (Summer 1979), pp. 353-355.

Haubitz (*Archaeopteryx* jaw – similar to other birds) *Paleobiology* 14 (2): 206 (1988).

Jerison, J.H., "Brain Evolution and Archaeopteryx" Nature, 219:1381-82, 1968.

Marsh, Othneil C., "On a New Class of Fossil Birds," *American Journal of Science and Arts* 5, no.2 (February 1873: 3).

Monastersky, R., "Hints of a Downy Dinosaur in China, *Science News*, vol.150 (October 26, 1996) p.260.

Monastersky, R., "Paleontologists Deplume Feathery Dinosaur," *Science News*, vol.151 (May 3, 1997): p.271.

(Archaeopteryx A Powerful Flyer) S.L. Olson and Alan Feduccia, *Nature* 278; 247 (1979).

Ostrom, John H., "Reply to 'Dinosaurs as Reptiles'", *Evolution*, vol. 28 (September 1974), p.493.

Ostrom, J.H. (1979) "Bird Flight; How Did It Begin?" *American Scientists*, 67:45-46, p.46.

Ridley, Mark, "Who Doubts Evolution?" *New Scientist*, vol.90 (June 25, 1981), p. 831.

Schwabe, Christian, "On the Validity of Molecular Evolution," *Trends in Biochemical Sciences* (July 1986).

Spetner, Lee, "Strange Case of Archaeopteryx 'Fraud'" *New Scientist*, vol. 105 (March 14, 1985), p.3.

Stanley, Steven M., "Macroevolution: Pattern Area Process", P.2 (San Francisco; W.H. Freeman and Col., 1979), 332 pp.

www.ingramcontent.com/pod-product-compliance
Lightning Source LLC
Chambersburg PA
CBHW031809190326
41518CB00006B/253